I0510634

Genes, Polymorphisms, and the Making of Societies

Genes, Polymorphisms, and the Making of Societies

A Genetic Perspective
of the Divergence between East and West

Hippokratis Kiaris, PhD

Universal-Publishers
Irvine • Boca Raton

Genes, Polymorphisms, and the Making of Societies:
A Genetic Perspective of the Divergence between East and West

Copyright © 2021 Hippokratis Kiaris. All rights reserved. No part of this publication may be reproduced, distributed, or transmitted in any form or by any means, including photocopying, recording, or other electronic or mechanical methods, without the prior written permission of the publisher, except in the case of brief quotations embodied in critical reviews and certain other noncommercial uses permitted by copyright law.

Universal Publishers, Inc.
Irvine • Boca Raton
USA • 2021
www.Universal-Publishers.com

ISBN: 978-1-62734-345-9 (pbk.)
ISBN: 978-1-62734-346-6 (ebk.)

For permission to photocopy or use material electronically from this work, please access www.copyright.com or contact the Copyright Clearance Center, Inc. (CCC) at 978-750-8400. CCC is a not-for-profit organization that provides licenses and registration for a variety of users. For organizations that have been granted a photocopy license by the CCC, a separate system of payments has been arranged.

Typeset by Medlar Publishing Solutions Pvt Ltd, India
Cover design by Ivan Popov

Library of Congress Cataloging-in-Publication Data

Names: Kiaris, Hippokratis, author.
Title: Genes, polymorphisms, and the making of societies : a genetic perspective
    of the divergence between east and west / Hippokratis Kiaris, PhD.
Description: Revised and extended edition. | Irvine : Universal Publishers, 2021. |
    Includes bibliographical references.
Identifiers: LCCN 2021021087 (print) | LCCN 2021021088 (ebook) |
    ISBN 9781627343459 (paperback) | ISBN 9781627343466 (ebook)
Subjects: LCSH: Human genetics--Social aspects. | Human genetics--Variation. |
    Human behavior. | Cultural evolution. | Genes--Social aspects. |
    Genetic polymorphisms--Social aspects.
Classification: LCC QH431 .K425 2012 (print) | LCC QH431 (ebook) |
    DDC 599.93/5--dc23
LC record available at https://lccn.loc.gov/2021021087
LC ebook record available at https://lccn.loc.gov/2021021088

To my wife Ioulia
To my daughter Frosini and my son Harris,

For their patience, criticism and inspiration

Contents

PART II

Personality Traits at the Population Level

PART III

Perspectives

Preface

In 2020, the world experienced the COVID-19 pandemic, the deadliest event of our lifetime. The development of a vaccine was announced in late 2020, but deaths will continue increasing, and it will take a few years until normality, as we knew it, returns.

As of December 2020, more than 1.5 million people died from COVID-19 worldwide. In the U.S. alone, the most advanced technologically and the world's wealthiest country, the death toll approached 300,000 cases. In China, on the other hand, in which the pandemic originated, COVID-19 related deaths were about 5,000. The difference is even more striking if numbers are expressed as fractions of the total population. In the United States, with approximately 330 million people, COVID-19 deaths accounted for about 1 in every 1,000 people. In China, however, with a population of 1.4 billion, deaths by COVID-19 were less than 1 in 300,000 people.

The two countries responded to COVID-19 differently. Restrictions in the mobility of the population from one region to another, business and school closures, accessibility and enforcement for testing and other measures were directly responsible for limiting the health-related damages in China and they are based on how stringent the imposed regulations are. At the same time, they reflect how willing people are to follow such mandates. One could argue that a regime that is more totalitarian can easily impose laws that limit individuals' freedom while a more liberal government cannot. Imagine what would happen in the U.S. if Congress passed a law, as early as January 2020, when the emergence of the epidemic became visible, forbidding people traveling freely from Georgia to South Carolina. But the actual question is deeper than that. Could a regime similar to that in China survive in the U.S. or Western Europe? Of course, there is North and South Korea, and until recently East and West Germany and the Soviet Union, which followed different paths. However, the discriminating factor has to do with the force applied to maintain the contrasting regulations, and we can easily see what happened when these forces were weakened, and the people were allowed to express themselves more freely.

To survive and to reflect their population's mentality, different regimes require a certain degree of acceptance. For example, we can see this based on how China and the United States handled the pandemic differently. We see that in the Eastern vs. Western cultures as well as how they developed over the years. We see that collectivism characterizes Easterners, while individualism describes the West; this goes beyond narrowly viewed political systems as we see them today. In Japan, for example, a technologically advanced and rich country in the Far East, or even in tiny Singapore, collectivistic values exist and are recognized to a degree much more prominent than in the various advanced countries of the Western world.

Behavioral traits such as altruism, exploratory seeking, risk taking, conscientiousness (the tendency for responsibility, hard-working, goal-directed activities and organization) show different intensity in Eastern and Western people and generally speaking, characterize today their societies, as they did in the past years. However, the real question is whether Easterners and Westerners are inherently programmed for collectivism and individualism, respectively.

Popular TV shows, frequently based on books that appeared during the last few years, revolve about a common theme, that of alternative history. *Fringe, The Man in the High Castle, The Plot Against America* are only some of the several examples. The protagonists can be historical or not figures, see themselves in an alternative world, in which foundational historical events occurred differently, "streaming" history to alternate paths. Yet, characters and society remain the same. Maybe the share of power has shifted from one group to another, yet, regardless of whether in the end the balance returns back to "normal," struggles were the same as we recognize them in today's "real" society. More importantly, the same social powers are still recognizable in this alternative Universe. Intuitively and instinctively, this is the response to my previous question: People have certain innate traits, so societies are likely hardwired for different cultures as well.

Being different is the cornerstone of this book without any attempt to imply that one is better than others. What is advantageous for one can be a disadvantage for the other. As people evolved and developed different cultures, they followed their different paths throughout history. But now, people are forced to live together. Information, goods, people, and even diseases travel faster than at any other time in history. Several different cultures have developed in the world that should be not be regarded as just the "in-between" of the East and the West. It is only because East and West, as I will later explain, epitomize all the differences, the comparisons become less complex and more typical.

The main goal of this book is to understand the true nature of our differences and how these evolved and developed throughout the history of our species. To do so, I will start from when the first humans appeared in Africa. Then, I will describe how they moved around the world to colonize the Earth and what types of genes they carried with them. I will also describe the typical Easterners and Westerners as well as how they and their cultures relate to the ancient Greeks and Chinese. I will conclude by providing some specific examples for some genes that are known to predispose for different personality traits, how they are distributed in populations worldwide, and how they may associate with the behavioral traits we recognize in the Western and Eastern cultures.

I would like to thank Professor Ryan Nichols (University of California, Fullerton) for his constructive criticism, critical reading and useful suggestions. Also, I would like to acknowledge the help of Dr Jeff Young, my publisher, for making this second edition possible. Special thanks also to my daughter Frosini, who studies Political Sciences at the Honors College of the University of South Carolina and very responsibly undertook the challenging task to edit her father's writings!

Preface of 1st Edition

I am neither a geneticist nor a historian or anthropologist. As a biologist by training, I have received some academic and formal training in genetics. My knowledge—that is, ignorance, actually—in history and cultural anthropology is only empirical and superficial. Therefore, by having this excuse, I hope that I will avoid the rigorous critique from the academic experts in these fields. To such experts, my thoughts may seem naive and oversimplistic, or that I have "re-invented the wheel"—which, intellectually, is even worse. With these reservations constantly accompanying my endeavor, I have continued writing this book.

The issues I attempt to explore all spin around one central question: Why the different populations around the world have developed different and distinct cultures that eventually led to different historical outcomes and different ways, according to which the corresponding societies have been organized, and, in general, distinct ways by which life has been viewed and perceived. These and other relevant questions are examined in view of the different frequencies in various genetic polymorphisms in genes affecting behavior. Furthermore, I attempt to focus on a comparative outline, both cultural and genetic, of peoples and populations from the two major cultural lines and civilizations that have appeared in human history and persist until today: the Eastern (Asian) and Western (European and American). In other words, these questions are reduced to why the Western line of thought has been dominated by Aristotle's *reason* and *logic*, while the Eastern line of thought has been dominated by Confucius's *harmony, collectivism, and context dependency*. The main idea of this book is that the presence of different genes in the corresponding people has actually dictated the acquisition of these distinct cultural and historical lines, and that an alternative outcome might have been unlikely. Based on current trends related to the globalization of cultures and economies, some predictions are finally being made on the development of human cultures and the potential future of human history.

I want to thank Joanne Asala for her great job in editing the text and making it comprehensive and Dr Jeff Young, my publisher, for his help and valuable suggestions during all stages of preparation of this book. Finally, I want to express my gratitude to my wife Ioulia for her constructive comments, critique and patience for the whole duration of this project.

PART I

Introduction

CHAPTER 1

The Concept

The central notion of this book is based on a very simple idea—so simple that it can be considered as self-evident. If the genetic make-up of individuals, affects—if not dictates— their behavior, then shouldn't this also affect collective decisions and actions, if examined at the level of groups of people that share certain genetic characteristics? Shouldn't people that are genetically similar among each other exhibit similar trends in their decisions that have affected their culture and history? Such groups of people, with a genetically distinct identity, can be considered entire nations or even what we call races and ethnic groups. No matter how stringent the definition of homogeneity is, especially genetic homogeneity, it is really arbitrary and quantitative. In any case, though, it involves groups of people that genetically are more uniform than other people that belong to other "racial" populations. Therefore, it is conceivable that history, at least the part of it that reflects the outcome of certain decisions and reactions of human individuals, is also affected by the genetic identity of the people involved. In other words, different people would have made different choices that, in turn, would have created a different outcome to their history. If we take this a step even further, then instead of history we can extrapolate to whole cultures, that more collectively can describe the various manifestations of human intelligence and provide the frame at which choices are being made.

These are all applicable at the various levels of organization of such groups, from families in which the genetic relations are so apparent, to the anthropological bands and tribes and races—notwithstanding that there is not a scientific consensus regarding what, exactly, the human races are or how many (Molnar, 2005). This term historically was defined by using a combination of both biological and socio-cultural criteria. Regardless of whether Asian people can be classified into five or fifty groups, whether or not they represent a distinct and single "race" or many different races, it is clear that they are in principle more "identical" among each other and distinct as regards their physical characteristics when compared to European people, and vice versa. This is due to the existence of several features among them that largely reflect (and are reflected to) their genetic identity. Not that all Europeans can pass as Europeans by looking only at their physical appearance, or Asians for Asians. Many cases exist of individuals with intermediate (or mixed) characteristics that point to the fact that there is a continuum in the intensity of these features. Furthermore, it tells that it is not a single or only a few, but rather a combination of several different features that is used to describe different people. Thus, there are abundant grey zones that do not allow drawing strict barriers between distinct populations. In addition, for certain features the geographical localization may be tighter than that of others increasing the complexity of how different "races" are defined.

Quantity, or abundance in one group versus another is the key. Take curly hair for example. Thicker hair is harder to become curly, which explains why Chinese have straight hair. Variations in genes such as FGFR2 and EDAR, which are found more commonly in Europeans than in Chinese, are responsible for this difference (Fugimoto et al., 2009). Of course, Chinese with wavy hair or Europeans with straight hair are not impossible, just not very common. Generally speaking, traits that manifest in lower frequencies in certain groups while they are more common in other groups of people, or features that are stronger in the one group and milder in the other contribute to the differences we recognize today in different populations. This can be due to the occasionally extended inter-breeding, the mixing up of genes of diverse populations, that could be inten-sified during specific historical periods between people of different ethnic groups. It could also be due to the fact that there is not a single genetic characteristic present in all people of the same population and absent from all others. Unless, of course, we are talking about a small, frequently iso-lated population. Take skin color as an example. Several genes contribute to the color of our skin, hair, and eyes that come in different "versions." These different versions result in the production of particular pigments at various levels, that in turn, determine our color complex (Sulem et al., 2007). Interestingly enough, Asians and Europeans present these versions at different frequencies, which is why East Asian blue-eyed, blond peo-ple are that uncommon (Table 1). Thus, a single (and objective) criterion to classify an individual as a member of a specific race does not exist.

Table 1. Frequency of alleles determining hair, eye and skin pigmentation in Europeans and East Asians

	Population frequency (%)	
	EUR	EAS
rs1805008 C	94	100
rs1540771 C	49	72
rs1042602 C	63	100
rs1393350 A	24	0
rs12896399 T	43	35
rs1667394 T	76	27
rs12821256 C	12	0

It is all a matter of frequencies, ratios, and intensities—but we'll come back to that later. We'll see that characteristics, such as the epicanthic fold or the double eyelid, are considered typical for East Asians and are usually accompanied by light skin color.

At the same time, though, we have also seen individuals that belong to Western populations who have considerably darker-than-average color skin types, accompanied by pronounced double eyelids, characteristics that are considered more "typical" for African and East Asian people respectively. Frequently, notwithstanding not exclusively, in the world of show-biz, such exceptions and deviations from the mainstream characteristics are more common than in the average population, which probably implies the attraction, the appeal these "minority" traits elicit—a fact that possesses apparent implications in providing certain mating advantages!

Notions related to the genetic classification and the eventual categorization of the various "races" emerge when traits are discussed. Those are of course wrong scientifically and are completely outside my intentions. Even when we subsequently describe examples of certain genetic features that superficially may be taken as disadvantages, we have to keep in mind that those should be judged as such only within the certain context that they have appeared and stabilized in a given population. Certain examples point to the fact that even disease-related genes, such as those responsible for the development of sickle cell anemia and Glucose-6-phosphate dehydrogenase (G6PD) deficiency, have an eventual role in conferring resistance against infectious diseases. Thus, what can be viewed as a disadvantage in the first instance is certainly an advantage at a different environment. Many more are the cases of genes that, while not responsible per se, can modulate the development or the severity of specific diseases and have a different prevalence among people of different origin.

Obviously, since even in cases that involve pathological conditions the distinction between "harmful" and "beneficial" is not clear-cut, when we talk about behavioral traits, the whole picture becomes even more complicated. It is much more difficult to classify a characteristic as purely advantageous or disadvantageous for the individual that bears it when we focus on characteristics that affect human behavior and personality. For example, novelty-seeking is a behavioral trait related to the tendency for increased risk-taking and exploratory excitability. This trait, historically, might have produced a positive influence in individuals, since it might have facilitated progress and advancement. It is noteworthy to mention that it has also been related, genetically and behaviorally, with increased incidences of drug addiction. Does this remind you of Western people (or people of European descent) and their civilizations? How about the observation that the specific polymorphism that is related to this trait is quite uncommon in Asians?

As we will see in subsequent chapters, recent complex analyses and genetic modeling suggest that these polymorphisms, such as those related to novelty-seeking, are likely associated with the migratory patterns of human populations, providing a direct hint on how genetics might affect the history of certain people. And not co-incidentally, speaking about the migratory pattern of behavior, the easier (or more efficient) adaptation into a new environment is intrinsically linked to novelty-seeking behavior. We will discuss all of these issues in much greater detail in subsequent chapters, along with other analogous traits. Thus, what is positive in a specific view can be negative in another. The level of complexity increases even more by the observation that certain genetic traits, depending on the exact conditions that are being studied, may affect a variety of behavioral trends and patterns, and what we see and record is actually the collective outcome of all these behavioral variables.

Another issue that may arise throughout this book is related to the concept of "free will." In that sense, genetically speaking, the unavoidable question is this: How free is our will if we are actually hardwired, or at least predisposed, against certain behaviors, choices, and reactions that differ among individuals? Genetics though provide just the frame.

How our specific responses will be formulated depends on several other factors as well, that collectively we classify under the wide term of environment. However, no matter what the correct answer may be and whether the actual balance is shifted towards nature or nurture, our only rational option in life, both as individual persons and as members of a larger group or community, is to keep on trying to extract the best out of what we have. Within this context, as individual persons we experience the option for a will that is really free.

Keeping these issues in mind, my whole point is that different people respond or are more likely to respond, differently against similar stimuli, and that these responses are likely more common among people that are more similar genetically. The latter is quite likely to occur with people that belong to a group or population with similar (or more homogenous) genetic imprint. Thus, if against the same stimulus or during an encounter, A people are more possible to elicit a type-K response while B people are likely to elicit a type-L response, then A and B people are likely to take consecutively different decisions through historical time: The A people will repeatedly respond with a K-type response while the B people respond with an L-type response. And, importantly, the genetically similar offspring of these people will continue to make similar decisions whenever they face similar challenges, thus exhibiting an apparent consistency in the building of their culture and norms. These decisions eventually will be reflected in their collective history. If, for example, this K-response is to retreat and negotiate when they deal with offensive actions, while the L-response involves confrontation and "fight back" decisions, then it is quite likely that A people will be less prone to warrior-type cultures than B people. We can also imagine that another genetically regulated trait exists that makes A people more co-operative than B people. It is conceivable in that case to expect that the A people will develop cultures and societies at which their individual members will exhibit increased interdependency than the B people. Thus, their cultures will be more "collective," when compared to those of the B people, who in turn will have a tendency toward individualistic cultures.

Imagine now, another hypothetical example in which populations consist of a mixture of A and B people at different ratios. Such is the case for all "real" populations in which genes come at different versions. Each has different frequencies in the various groups of people. To properly operate, societies need both leaders and followers, novelty seekers, and those

7

who are reluctant to change. Additionally, they require individuals who relentlessly follow instructions and routines, as well as rebels who do not. Instead, when the entire process becomes dysfunctional, they are the ones who think innovatively and provide resolutions. Kastoriadis (1964)[1] suggests this when describing the worker's role in production of goods but can be readily applied to the operation of the society as a whole as well: *"... the worker experiences the absurdity of a system seeking to turn him into an automaton, but obliged to call on his inventiveness and initiative to correct its own mistakes"*. However, what are the consequences of having people with these personality traits in different ratios in society? The complexity increases, even more, when we consider that it is actually a combination of traits that produce the outcomes of potential interest.

Of course, the unbiased question is whether such genetically regulated traits exist that can affect those types of cultural and historical decisions. This is a main focus of this book, and we will try to address it as we move forward.

Consistently with these, it is not only the socio-economical environment, the geography, the natural phenomena, the occurrence of certain disasters and diseases or other exogenous factors that have affected and will continue to affect the history of humans, but also the genetic signature of the people. Therefore, in any attempt to explain human history, the genetic profile of the corresponding people should also be taken into consideration, along with the other conventional, exogenous factors. Even if this is not feasible technically as yet, it is rather likely that in the near future and to a certain extent will be. And in that case, we can even go one step further; besides explaining the past, we might also be able to "predict" the future. This may sound like a science-fiction scenario right now, but if we were able to "measure" behavioral tendencies and the genetic structures of given societies, then the prediction of possible outcomes against specific conditions could be made. Since retrospectively we can explain various outcomes of history, then why not be able to predict them? The "complexity" factor is, of course, a chief parameter. Yet, the chaos theory in mathematics tries to do precisely that. It tries to explain events that appear as random but eventually follow specific rules and can be subjected to modelling.

That this "runs in the family" does not only refer to diseases or certain physical features, but also to behaviors, our likes and dislikes that, conventionally, we used to attribute only to certain socio-environmental factors—in other words, to the way we "grew up." For example, that ancient

Greeks (or, to use a present-day example, the Kennedy family) were deeply political may reflect, at least in part, their genetic signature. A certain predisposition against specific behaviors, such as the tendency of not taking things and conditions as a given, in combination with the desire to lead, may manifest as an attraction to politics. Or to science. You need to feel that something is not "right", or adequately explained to be able and build into that and make a change.

Not that the environment—in its widest sense—is not playing a major role, of course. On the contrary, actually! However, in order to formulate characters and mentalities (as well as physical characteristics and diseases), the environment needs to interact with the given genetic background, and the result will be as specific as it can be for each individual. Thus, the behavioral pattern that emerges is not the sole result of the environment, but is also greatly affected by the genetic information carried by each and every person as well. This is, of course, a given, a basic and elementary knowledge of biologists. The notion of context dependency ranges from the study of the genetic basis of behavioral traits, and it affects the interpretations of those attempting to understand the response of individuals against certain stimuli, to the elucidation of the effects of the (micro)-environment in cellular differentiation and disease. Or to why the same carcinogens cause cancer in some individuals but not in others. It is increasingly appreciated that the magnitude and the extent of any biological response is greatly influenced by the genes of the individual.

In a reductionist's approach on addressing how genes affect behavior, the emphasis is given to the prediction of how individuals will respond to a specific (socio-economical) environment, or in other terms, to the building up of "characters" and personalities in their wider sense. It is reasonable, though, to speculate that analogous mechanisms will also operate at a larger scale as well, at the level of populations, and in that case, the outcome will be reflected not to individuals' decisions, but rather to collective decisions that are capable of affecting history. By extrapolating from these ideas and projecting them to whole populations, we can probably start to understand choices that may appear random. We can explain why certain paths were preferred over others, at least by certain people as compared to different groups of people. This way, we may understand why China was very successful in dealing with the containment of COVID-19 over the West. We may be able and appreciate why during its history, Europe was and still is fragmented to various countries with people feeling highly different than their fellow Europeans, while China is characterized by a great sense of stability and continuity during its history. We may also appreciate why Aristotle was born and taught in Greece while Confucius left its impact in China.

We can read in J. M. Roberts's classic book *A Short History of the World* (1997) that "[H]uman history began when the inheritance of genetics and behavior which had until then provided the only means of survival was first broken through by conscious choice" and "[I]t (human culture) was increasingly built by deliberate selection..." (p. 2). The question, in that case, is related to how strictly we define the terms "broken through by conscious choice" as well as "deliberate," and on whether all conceivable options indeed carry the same probability for different groups of people...

CHAPTER 2

Genes, Polymorphisms, and Genetic Heterogeneity

Individuals are different. They differ in their physical appearance, their specific abilities, their inherent risk to get sick from certain diseases, their characters, and their entire existence. Depending on the particular characteristics we are studying, such differences can be attributed, to some degree, to the individual's DNA, or in other words, to the genes that someone has inherited from his or her mother and father. We are all familiar with examples of sisters that one is very extravert, popular at school and social, while the other is the "stay at home" type, seldomly goes to parties and her friends can be counted in the fingers of a single hand. Alternatively, we are also aware of "like father like son" examples at which the son is a personality or appearance replica of his father.

For some characteristics, the contribution of the genes is absolute and apparent, while for others, the environment may play an equal or even more important role than that of the genes. Thus, while our eye color is solely determined by our genes, our character and behavior are also determined by the conditions we grew up in, our friends, our family, and, in general by the environment, or alternatively, by nurture.[2] Again, even in the case of traits that are so obviously shaped by socio-environmental factors, a not negligible genetic contribution is there. Savings behavior, for example, a trait that is thought to be affected grossly by the environment, it is increasingly appreciated that it possesses a substantial genetic component (Cronqvist & Siegel, 2010). Interestingly, according to this study, the strength of the genetic expression of savings behavior is modulated by whether the subject grew up in a supportive environment, exemplifying the interplay between "nature" and "nurture" in a behavioral trait.[2] More importantly, opposite or complementary trends are needed in a society to function well. Imagine now the implications of having a group of people or even a population that has the genetic inclination for savings at more prominent levels than that of another group that prefers spending over saving. The "savers" group would not allow enough money to circulate since consumption is restricted, and the economy will be stuck. In the opposite case of the "spenders" society, there would be no planning, and no back-up funds will be available to sustain them in occasions of potential crises. Both would eventually collapse. Financial planning suggests modifying our strategy accordingly towards the spenders or savers style in our lifetime to maximize our goals. However, both savers and spenders are needed, even if they are the same individuals during different life stages.

What are these genetic polymorphisms that underline our individual identity? Let's turn to some basic concepts. Simply speaking, the term refers to the existence of different "versions" of the same genes that, while in principle are identical, exhibit a slightly different activity which eventually is interpreted to changes in their functionality. This is due to slight differences in their structure (primary sequence) or their regulatory regions. The differences can be quantitative (for example, stronger or weaker activity of some enzymes) or qualitative (such as a product with different color). It is noted, though, that most of the time, these qualitative differences can be reduced to differences in the activity (quantity) or the amount of the pigment for color-related phenotypes, but for our purposes this doesn't make any actual difference.

The variation between individuals in virtually all their physical characteristics is due to such polymorphisms, or in other words, due to the fact that different persons carry different versions (and the combinations) of the genes. This is why two brothers can have different color of hair, two sisters have different skin types, or a father has wet-type earwax while his son has the dry type. It's just because they have inherited different versions of the genes that control these traits. Given that humans have about 22,000 genes, and that many of them are polymorphic, the combinations are practically unlimited. Furthermore, many genes have more than two polymorphic alleles, a fact that increases the possibilities even more. This is the genetic basis of human variation, or vice versa, human individuality.

The reason this variation exists is usually related to the evolutionary and adaptation processes, as well as to phenomena associated to pure randomness. In many cases, the benefits conferred by certain alleles and the resulting characteristics are apparent. For example, in human populations, darker colors in the skin confer increased protection against harmful sunrays. This is the reason that darker skin colors are more common in populations closer to the equator, since people in these places are exposed to more time in the sun. On the other hand, since sunlight is also important for the metabolism of vitamin D, people who have darker skin color (due to increased amounts of melatonin) in geographic areas with little sunlight, such as in the northern climates, often have deficiencies in vitamin D. So, what is an advantage under certain conditions can be a disadvantage under different conditions. Obviously, the specific environmental conditions play an important role in attributing a certain genetic feature with a beneficial or negative character.

In other cases, the advantages are more obscure and hard to identify, such as in the case of the earwax type, for which the dry-type has been proposed to be an adaptation against colder climates. Even the variability in a "minor" characteristic, such as earwax type, can produce benefits affecting the domination of one population over another, reflecting the cumulative power and magnitude that even slight genetic variation may have in whole populations. Of course, new alleles, or different versions of the same genes, do not necessarily need to offer selective advantages in order to be stabilized in a population. Even more importantly, these new genetic variants do not necessarily need to offer such advantages instantly, at the time they appeared in the individuals. They may just be present in the population, increasing the variability and waiting for the conditions to change so natural selection and evolution can utilize them. Thus, it is not only natural selection that drives evolution. Other complex phenomena, such as genetic drift, may account for this stabilization. Genetic drift essentially incorporates random phenomena that affect the frequency of alleles and are stronger in smaller populations. So, if a new genetic variant appears suddenly in a small and relatively isolated population, it may be stabilized and eventually dominate this population, even if it doesn't affect the individual's survival. This variant does not always have to appear from "scratch" in a single individual. It may be present in the larger population already at some low frequency. If however some individuals that have this rare allele decide to migrate, in the new population that they will establish the rare allele will not be rare any more (Ridley, 2003; Futuyma, 2009).

By an analogous manner, we know that certain diseases are controlled by specific "sick" versions of genes (more accurately speaking, genes encoding for products related to sickness or disease), while others regulate our predisposition or modulate our chances (or risk) to get a disease. In other words, their presence operates as a genetic risk factor for certain conditions. The stabilization of these "sick" genes within a population occasionally may also be associated to an advantage contributed by these genes in a certain environmental context. In sickle cell anemia, for example, the allele that is related to the disease, when it is found in heterozygosity (one copy of the sick gene and one copy of the normal or wild type allele in diploid organisms, such as humans), renders individuals more resistant to malaria, and this is the reason it is more common in geographical areas with higher malaria incidence. This is an example of balancing selection that favors the presence of multiple alleles within a population as opposed to directional selection that favors only a single allele. In other cases, disease-related genes may be stabilized and prevail in certain populations due to cultural and other phenomena that are not related directly to the genes' function. One such example is the relatively high prevalence of Usher syndrome in Samaritans that causes congenital deafness. Samaritans are related genetically to Jews in terms of being descendants of a group of Israelite inhabitants. While in ancient times they exceeded a population of one million, today there are just a few hundred, and they are considered as one of the more highly inbred populations of humans. Due to the high degree of inbreeding and their relatively low population numbers, a specific mutation in chromosome 11 that causes Usher syndrome has been stabilized in this population. Thus, individuals that have inherited two copies of this gene from their parents suffer from deafness (Bonné-Tamir et al., 1997).

Whenever an apparent association does not exist between the presence of certain "disease"-related alleles and specific advantages, while cultural and historical parameters cannot explain their presence, the reasons that account for these alleles not having been eliminated from the general population are under debate. Sometimes they may reflect advantages that we have not discovered as yet. In other instances, they may have just happened and only time will tell what their impact will be. In any case, we have to keep constantly in our minds that natural selection, by definition, is a dynamic and constant process that never ends, and what we see now is only a snapshot of evolution. So, in other words, what we see now is not an evolutionary end-point at which the selection processes have been completed and the best possible outcome produced. There is always the possibility that what we see as a negative trait today will be eliminated in the future or turn out to be advantageous when the conditions change.

Or new mutations may appear in specific genes that, due to an advantage they offer, the individuals bearing them may increase in numbers and the corresponding alleles may be stabilized in the population. In that latter case, we may even witness that other alleles located in the vicinity of these advantageous alleles may also be favored, indirectly.

While the whole situation, in terms of genetic contribution, is relatively clear, as regards some well-defined physical conditions and characteristics, in discussing behavioral traits the landscape is by far more elusive, and thus the corresponding scientific controversies more vivid. The explanation is simple, and it is related to both the complexity of these characteristics in terms of how the corresponding genetic loci control the development of behavioral traits, as well as the not precisely understood, and indeed defined, contribution of the environment to behavior. Simply speaking, it is easier to identify the contribution of nutrition as an environmental factor in regulating, besides his genes, someone's height (or for a smoker with a history for lung cancer, the possibility to get the disease) in the presence or absence of specific polymorphic alleles, than understand whether some violent behavior, besides the environment, is also due to the genetics of the individual. In addition, personality traits at the level of society operate in concert to produce their outcomes collectively. Consider the spenders-savers example discussed above. However, now add a personality trait, such as conscientiousness, which describes the tendency for good performance at goal-directed activities. This suits the savers better and can be a prerequisite for them in order to engage in planning. Therefore, all these that can be the long-term beneficial consequences of the "saving" behavior. So, such traits work better when they are combined with other personality traits, which, in turn, suggests that groups of personality traits co-exist to provide beneficial outcomes.

Studying behavioral traits has additional difficulties. Speaking about how society, in terms of a given cultural context, might judge and classify behaviors and behavioral traits, are not static notions, but are rather dynamic and are highly influenced by the socio-cultural environment and the historical time-point we refer to. These notions are illustrated very nicely by studying how different and occasionally extreme behaviors were considered (or not) to be pathological in different historical periods. Such issues are discussed, among others, in high detail and depth by Michel Foucault in his classic *History of Madness* (2006). Thus, if even the borderline between normal and pathological is not strict and objective, then it is understandable why the classification of an individual's trait A as A1 or A2 type is also inconclusive at the least. On top of those, experimental animals can be used to study cancer and diabetes but not for studying savings behavior.

Despite those—conceptual or not—limitations, a lot of progress has been made and various behavioral traits have been linked to specific genetic variations. Or, vice versa, certain polymorphisms have been linked to an increased probability against specific behaviors. The term "probability" is of particular importance because it indicates that merely having the gene doesn't mean that someone will develop the trait, but rather that he has higher or lower chances to do so. Or, alternatively, among a group of people that carry the gene, only a fraction will develop the trait. These "chances" (reflected to the genetic penetrance of the trait) among others are determined by the interaction of the gene with the environment. In other cases, they are attributed to pure randomness or stochastic phenomena. Furthermore, the chances also change by the activity of other genes that can also be polymorphic producing effects that for geneticists are known as "epistasis".

Naturally, the precise manifestation of these behaviors, their degree or level of expression, and their exact type are formulated by environmental factors, as well, but undoubtedly, the genes we carry also play a major role in the determination of personality. In several cases, a link between various traits and specific genetic loci has been established with some being better supported by the experimental evidence than others. Those include, but are not limited to, behaviors such as the exploratory activity, novelty-seeking, tendency for aggression, social behavior, or even risk-taking behavior related to financial decisions. Genetic variations in (brain-) hormones, neurotransmitters, their corresponding receptors, and other proteins related to their metabolism or physiology, are the usual suspects that may explain the differences.

The role of neurohormones in the decision process has recently been described, among several other interesting examples, by Lehrer in his book *How We Decide* (2009). This observation is not surprising, considering the pleiotropic action of such hormones and their involvement in determining our behavior and responses. An example is offered by the polymorphisms in the serotonin transporter gene. Serotonin, or 5-hydroxytryptamine (5-HT), is a neurotransmitter that is derived by the modification of the amino acid tryptophan. Amino acids are the building blocks of proteins; tryptophan is one among 20 of those. Among other tissues, tryptophan can be found in the central nervous system. It is acting via specific receptors that, in turn, can affect the activity of other neurotransmitters, such as epinephrine, dopamine, and achetylcholine. Serotonin levels are regulated depending on specific internal and external stimuli, and have been associated with various behaviors and conditions, such as appetite, mood, emotion, and sleep. Proper activity of serotonin is facilitated by serotonin transporter (5-HTTLPR), a protein responsible for the re-uptake of serotonin from the synaptic cleft. Importantly, the gene encoding for this transporter protein is polymorphic. A certain variation in the promoter region of this gene that corresponds to the gene's regulatory region, which determines when it will be activated, results in two versions of the gene: a long version (*l*) and a short version (*s*), with the latter associated with reduced transcriptional activity and, thus, decreased availability of the serotonin transporter. As regards a potential link of 5-HTTLRP and behavior, the long allele has been associated, among other traits, with "happiness," while the short allele with anxiety and impulsivity. It's worth noting that the short version of 5-HTTLRP is found in Asian populations almost twice as much as in Caucasians, with considerable variation among Europeans. Based on this, in a linear projection, one would expect that Asians are more impulsive and anxious than populations of European ancestry. Yet, the actual impact of this variation at the population level is thought to make societies more protective and structured in a way that makes them able to provide the psychological support that is needed when the genes of the individuals cannot. This, in other words, may promote the establishment of adherent links among the people of these societies. Such links are dictated by the genes, but only indirectly, as a countermeasure against some other undesirable effects. Yet, the results are still the same.

This book attempts to explore various selected behavioral traits in view of their potential contribution in shaping cultures, history, and eventually historical decisions. The genetics of these traits will also be discussed in light of the prevalence of specific polymorphisms in different ethnic groups and how such differences may account for the different characteristics between people of different races, or how they might have affected their culture and history.

The study of various polymorphisms in different human populations is also of great interest to anthropologists; however, such studies usually focus on the distribution of these polymorphisms. By using them as markers, biological anthropologists attempt to identify similarities or divergences between different populations, patterns of migration and intermixing, and other factors related to those characteristics. Besides their function, such "meaningful" polymorphisms may also exhibit occasionally a specific contribution to behavior and may affect differentially some general trends or traits in various distinct populations. This latter notion is the main focus of this book. We will concentrate particularly on the "comparison" between people of Asian and Western origin, since these people exemplify the two major cultural lines of humankind. By analyzing cultural trends and characteristics of Eastern as opposed to Western civilization, we will explore whether they are in line with specific genetic trends described for these people, which in turn may indicate the existence of a causal association between genes and cultures.

2.1. Genetic Markers and Analysis

When talking about polymorphisms and different versions of our DNA, it has to be mentioned that it is not all our DNA that corresponds to genes (that, in turn, and as mentioned before, comes into different versions). Actually, the vast portion of our DNA, at the magnitude of about 98%, is "junk." It doesn't have any specific function (as far as we know), but it's still there. And it exhibits even stronger variability, or polymorphic incidence, than the coding DNA between individuals. This is not surprising, considering that it isn't the subject of evolutionary pressure. If something happens to this, no imminent consequences will be observed in the cells because it does not encode for a protein.

A major—and genetically useful—source of variation in the DNA corresponds to small changes in the repetition number of specific core, repetitive sequences. Most of the times, such changes in repetition numbers are meaningless, while in others, such as when they are located in chromosomal regions that regulate gene expression, they may have consequences as they can change how actively a particular protein is made.

Another source of variation is associated with base substitutions (that occasionally, if they are found within the gene's coding sequence, don't change the corresponding protein). The onset of a mutation in a protein-coding segment of DNA can quite possibly affect the cell's—and, by extrapolation, the individual's—viability. Whether a certain protein exists in versions with higher or lower activity, due to polymorphisms in its genomic sequence, it is quite possible to affect some physiological process, with the one version being "better" than the other, at least within a given environment. Most of the time, these mutations are eliminated because pure chance in their occurrence is more likely to generate "bad" alleles. Infrequently, these random genetic events are associated with the contribution of a beneficial feature, a sort of advantage, and eventually they will dominate the population if they contribute to the individual's (mating at the end) performance, or the performance of the group as an entity (in cases of group selection). This is the essence of evolution and refers to the cases at which the changes interfere with some kind of activity or overall efficiency.[3] On the other hand, if the mutation has targeted a non-coding segment of our genome, namely our "junk" DNA, it is not going to affect the individual's viability. In these cases, and by mechanisms that do not directly involve selective processes, they can be stabilized in the population and passed into the subsequent generations by following of "unbiased" inheritance. The term "unbiased," in this case, refers to the fact that the chances of inheritance into the subsequent generations of any of the two potential alleles are equal. This is the case with most of these non-coding polymorphisms, unless they are close to other coding sequences that also exhibit variability. In that case, the non-coding polymorphisms "benefit" from the selection that operates onto these coding polymorphisms and may increase (or decrease) in frequency in a population. Thus, they are benefitted by their good (or bad) neighbors and their abundance will eventually reflect the abundance of their neighboring genes. Not infrequently, geneticists identify a specific polymorphism that is associated with a particular genetic trait and try to understand if this polymorphism bears a functional significance or if it is just close to other polymorphisms that carry this functional significance.

The absence of evolutionary pressure in these sequences renders the corresponding polymorphisms ideal genetic markers (Griffiths et al., 2000). This is, among others, of particular importance in tracing the fate and history of DNA sequences and is also useful in order to identify the physical location of genes that are associated with certain traits and diseases. If, for example, a certain polymorphism is seen in populations A and B but not in C, it is quite reasonable to assume that A and B people are somehow more closely related among each other than with the C people. Now, if we also take into consideration their geographic location, our conclusions can also be extended to migration events and the history of these populations. Analogous and even more refined conclusions can be obtained if we take into consideration, instead of their mere presence or absence, the

19

frequencies of these polymorphisms in the corresponding populations, as well. In subsequent chapters, we will describe the results of such analyses and how, for example, they offered us clues regarding the migratory events that resulted in the colonization of the European continent and the rest of the world. If, on the other hand, among the same population a specific polymorphism is found more frequently among people with a certain trait (or disease), then it is quite reasonable to conclude that either causally (as part of a coding or regulatory DNA sequence) or due to its close proximity with a functional genetic variation, this polymorphism is associated with the onset of the corresponding trait.

Of use in genetic analyses, particularly those related to anthropological studies, is mitochondrial DNA. The mitochondrion is a cellular organelle that works as the cell's power plant. Noteworthy, it comes with its own DNA, a short, circular, 16,000-base pair long DNA segment that, surprisingly, in its structure bears greater similarity to the DNA of some bacteria than with the cell's chromosomal DNA. Actually, it is considered that mitochondria trace their cellular ancestors to bacteria that "decided" during a certain evolutionary step to live together with other cells that are the ancestors of eukaryotic cells, quitting permanently their individual identity. A similar event also happened with the plant cells and their chloroplasts, the organelles that perform photosynthesis.

An important aspect in their biology, besides their obscure prokaryotic-like structure, is their maternal mode of inheritance. Namely, all of our mitochondrial DNA comes exclusively from our mothers, who, in turn, received it from their mothers, and so on. Interestingly, even this mitochondrial DNA is polymorphic, a fact that renders it particularly useful and informative when we examine maternal lines of genetic inheritance. Thus, by tracing the polymorphisms in our mothers, we may track down our maternal ancestors. Indeed, as it will be discussed in greater detail in the next chapter, the concept of our maternal ancestor, who has been called the "Black" or "Mitochondrial" Eve, has originally been formulated and supported by studies of polymorphisms utilizing mitochondrial DNA samples obtained from different populations.

CHAPTER 3

Biological Anthropology and the Distribution of Human Populations As We Know Them Today

*H*omo sapiens[4] acquired its formal name during the eighteenth century from the Swedish botanist and zoologist Carl Linnaeus. By observing the differences in the physical characteristics among individuals, he went a step further, dividing humans into four subspecies: *Homo sapiens europaeus, Homo sapiens americanus, Homo sapiens asiaticus,* and *Homo sapiens afer* or *africanus* (Marks, 1995). Color played a major role in this classification, corresponding to white, red, yellow, and black skin, respectively. His conviction that this classification was extended beyond skin color was so strong that he attributed certain behavioral characteristics to each of these taxa. So, according to Linnaeus, the individuals that were classified as *Homo sapiens afer* were relaxed, crafty, and negligent, while those of the *Asiaticus* taxon were avaricious and distracted easily. Native Americans, or *Homo sapiens americanus,* were stubborn, merry, and easily angered, while *Homo sapiens europaeus* were gentle, active, smart, and inventive. Finally, he also proposed another variation within humans, termed *Homo sapiens monstrosus,*[5] which include people with deformities, relatively unknown groups such as the Khoikhoi people of South Africa, as well as other fantastic people (de Vaal Malefijt 1968).

While, of course, such classifications are both ethically unacceptable and scientifically wrong, among others they signify the fact that a high degree of variability exists between human beings and that this variability is not random but is rather associated with a specific geographical distribution. In other words, it is more likely to find black people in Africa than in Northern Europe and people with double eyelids in East Asia than in the Mediterranean. Furthermore, it implies that certain behavioral traits might also be associated with specific populations. Of course, the presence of certain features is neither necessary nor sufficient to derive conclusions regarding the geographic origin of individuals. It is just correlative and denotes what is more likely to be than what actually is. And this is exactly the "tricky" part in all interpretations: While we may be able and identify behavioral patterns in different populations, we are always also able to point to exceptions that cast doubt in all generalizations we are attempting to make. In addition, almost always the environment plays significant roles on how such patterns are expressed and therefore we tend to attribute the differences we see in the context rather than in the biology.

Nevertheless, people are different and thus, their populations are different as well. First, let's try to see how all this variation between people developed and why it is related to geography. To do that we need to identify it origins. In order to trace the source of this variability among human beings and its distribution, we have to travel back a few million years, to the time when the earliest ancestors of the evolutionary line that ultimately resulted in the appearance of *Homo sapiens* can be identified. Technological developments, new findings, and alternative interpretations of existing experimental data constantly appear in the scientific literature that continuously change the details of this story; however, the outline is not distal from what I will describe below. Olson (2002) and Sykes (2002), in their corresponding books, provide very well-written accounts of the processes that resulted in the appearance of modern humans and the distribution of human populations around the world.

Such a trip will take us back around four million years, back to when the most "advanced" species, in evolutionary terms, were the apes. At this time, certain apes (that today have been classified as belonging to the genus *Australopithecus*) have acquired the ability to stand on two feet. This transition into a standing, or two-footed, life is considered a major breakthrough and is accompanied by a quite important domino effect of changes initiated by the alternate use of the front feet, now hands, to handle and eventually manipulate objects.[6] We should not forget that human civilization is intrinsically linked to the manufacturing of certain tools.

Australopithecus was quite successful since lasted for about 2 million years. The first species that has been classified as belonging in the genus *Homo* (the same as today's modern humans) appeared two million years ago. Species of the *Homo* genus had larger brains, an indirect consequence of standing on two feet, and, occasionally, a remarkable ability to make, hold, and use primitive objects. This ability progressively improved.

The earliest *Homo* that could be classified as human, or *Homo sapiens*, appeared around 100,000 to 200,000 years ago. Consistent with this view, which represents our current understanding regarding our evolutionary history, we are descendants of *Australopithecus* through several intermediate steps of speciation. How do we view human speciation? Until recently, gradual changes in pre-existing species were thought to play a major role in the onset of a new species; such notions were progressively abolished and consecutive speciation events are considered to be the principal mechanism of human evolution. In other words, a subpopulation of one species was becoming isolated and was then giving rise to the derivative species. This isolation can be, for example, geographical or reproductive, and is not necessarily absolute and firm, or by speaking with qualitative terms "black and white," or genetically by being able to breed or not able to breed with other individuals outside the population. Reduced efficiency in the interbreeding between individuals of such isolated populations usually may be sufficient to attain isolation. So, if offspring numbers are reduced by 50 or 60% between two groups within the same population, then this may be sufficient to result, in evolutionary terms, in the generation of isolated sub-populations and the onset of a new species from a pre-existing one at some time.

Naturally, these evolutionarily related species may co-exist for extended periods of time. This was the case for certain populations of the genus *Australopithecus* and populations of the genus *Homo* around two million years ago, or the Neanderthals and the modern humans until 30,000 years ago. However, one species in the end dominates and the other species eventually disappears. This is the case with modern humans, or *Homo sapiens*; their appearance about 200,000 years ago resulted in the extinction of all other related species, including the Neanderthals, that today we may call archaic humans.

Our closest relative today is the chimpanzee. It is estimated that our common evolutionary ancestor lived about six million years ago, at a time when a population of this primitive ape split in two, one leading to chimpanzees and the other to modern humans. All intermediate species have disappeared, with Neanderthals being the closest archaic human that co-existed with us chronologically, until about 30,000 years ago. So, for 100,000 years or more, humans and Neanderthals co-existed. At the beginning, the former were represented only by a tiny population that was living

23

in Africa, while the latter were more widely spread. Progressively, though, the number of humans increased, and that of Neanderthals decreased, and by 30,000 years ago, only humans could be found—and we soon dominated the whole globe.

Until recently, and by data based predominantly in the fossil records, anthropologists thought each major population of modern humans was derived from a distinct species of archaic humans that lived in the corresponding geographical area. Consistent with this notion, Asians descend from *Homo erectus,* since the latter was living predominantly in Asia; Europeans come from the Neanderthals that lived in Europe; and Africans from other archaic forms of *Homo sapiens.* According to this model, which is called multiregionalism, it is the *Homo erectus* that principally migrated out of Africa as early as two million years ago, and modern humans emerged throughout where archaic humans lived (Figure 1).

The multiregionalism theory, while it still has several proponents, progressively has been replaced by the "out-of-Africa" model. (Figure 1) This theory, which currently represents the most widely accepted view on human evolution, posits that modern humans, namely *Homo sapiens*, evolved quite recently, about 200,000 years ago in Africa, and subsequently migrated throughout the world. The first major attempt for human migration (we think) occurred around 100,000 years ago and continued

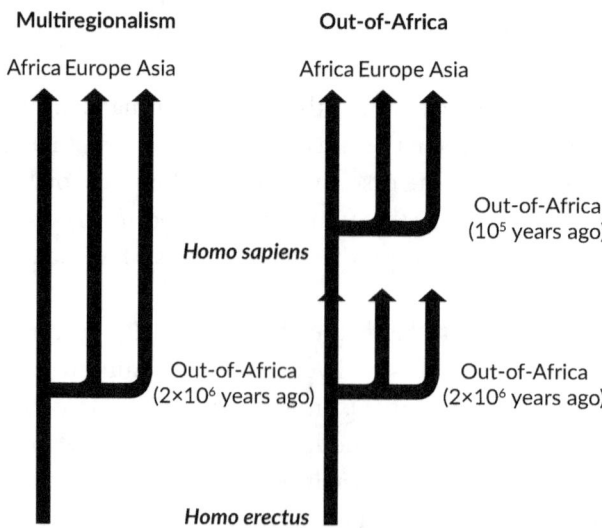

Figure 1. Diagrammatical depiction of the multiregionalism and Out-of-Africa theories.

repeatedly after that. Whenever populations of *Homo sapiens* encountered a population of archaic humans, the former replaced the latter, with Neanderthals being the last archaic humans replaced by modern humans in Europe, around 30,000 ago, as evidenced by the age of the most recent findings.

Neanderthals caused a lot of debate regarding their evolution. It has been proposed that they represent a different species, namely *Homo neanderthalensis,* or a subspecies within the human species named *Homo sapiens neanderthalensis.* We know that, although with limited efficiency, Neanderthals could interbreed with modern humans and exchange genetic material. They had also developed quite earlier than modern humans, around 500,000 years ago, and lived in Europe and Southwestern Asia until their extinction. Several theories have been proposed regarding the disappearance of the Neanderthals, such as weather-related explanations, shortage of food, or even Creutzfeldt–Jakob (mad-cow) disease associated with the fact that they may have been practicing cannibalism; however, the true causes remain elusive.

3.1. Humans Around the World Are More Similar Than They Are Different

It is estimated that the first population of *Homo sapiens*, around 200,000 years ago, consisted of about 30,000 individuals who lived in Eastern Africa. Indicative of the fact that all modern humans share a recent common ancestor is that, at the genetic level, humans show little geographic variation, as compared to other species. Various approaches have been applied to show the degree of genetic variation. A simple approach is the following:

In a given population and for a gene at which two versions (alleles) exist (A and a, respectively), heterozygosity (H) corresponds to $H = 1 - (p^2+q^2)$ and reflects the possibility to get the two different alleles together. Now, if more than a single population exists, then genetic variation $G_{st} = (H_T-H_S)/H_T$. In this case, S is a subpopulation while T is the total population. By taking a closer look at this equation, we'll see that values for G_{ST} range between 1 and 0 for the cases when two populations show completely different or identical frequencies in the alleles present. It has been calculated that among different human populations, G_{ST} values are relatively low, at the range of 0.07, while in other species it is generally higher, such as 0.12 for mice and 0.67 for animals such as Kangaroo rats. In an analogous approach, it has been calculated that in humans about 85% of genetic variation occurs within

25

groups and only 15% of variation between groups. Thus, no matter what differences can be identified, humans from different populations, let's say from different continents, are more similar among each other than they are different. This is quite important in order to place in the correct perspective the differences we will discuss later regarding people who live in different geographical locations and belong to different populations. On the other hand, elephants in Western and Eastern Africa display genetic variation of about 40%. The cut-off for subspecies is considered 25–30%. Thus, these elephants can be considered as different sub-species, but not the humans that belong to different "races." Humans are quite similar genetically and not sufficiently isolated, since the time of their relatively recent development. In that view, a difference of 15% in humans between groups is small. The question, of course, is in what sense we can define "big" and "small" in these issues, and whether 15% can make a difference. We have to consider that, as we'll see in subsequent chapters, certain polymorphisms at specific genes result in differences in activity at the range of 10–20%. These differences alone, assuming that they are the only differences among the individuals, are considered sufficient to cause a predisposition against certain behaviors and, in the widest sense, phenotypes. Furthermore, we also need to consider that such behavioral and other traits may usually "go together" in different population groups.[7] In that sense, we frequently see instead of a single trait, a set of traits that co-exist in different populations. Furthermore, these traits occasionally exhibit a complementary—which makes the archetypical Western-like and Eastern-like behaviors, and eventually cultural norms, even more robust and distinct. Whether now this "complementarity" provides advantages to the populations or it just happened to be like that is debatable, but as we will see it did happen and appears to make sense retrospectively.

As an example of a non-behavioral trait (at least formally speaking, since it may be intimately linked to dairy production and consumption in Japan), we can mention that Japanese are, in their majority, lactase deficient. They frequently also have epicanthic folds. So, these two characteristics usually go together. Assuming that these two traits (lactase deficiency and epicanthic fold) were related to behavioral patterns, and when they interacted or were combined they could result in the onset of more potent (and occasionally different) characteristics, then we could understand that what we record in behavioral terms is unavoidably the result of certain synergies. Finally, by referring to populations in the context of historical time, and not just the time that a behavioral experiment lasts or even the life of a human individual lasts, we may postulate that these effects will be cumulative and likely multiply progressively in the generations of people. Think of memes as an example.[8] In this case we will have memes potentiated by genes as well!

3.2. The Contribution of Mitochondrial DNA-related Studies

For our understanding of the human origins and the geographical distribution of people around the world, mitochondrial DNA analyses played a pivotal role. Mitochondrial DNA, since it is transmitted almost exclusively from mother to daughter, is particularly useful in tracing maternal lineages. Its inheritance is more "linear" than that of chromosomal DNA and the information it provides is easier to interpret. Furthermore, from the degree of variation, we may calculate the number of generations, and thus obtain clues regarding the time that has passed. In that sense, it may function as an evolutionary clock. If we think of DNA as a watch, minutes will be shown by the chromosomal DNA while the hours from the mitochondrial DNA. In order to understand this, let's think about the following: In each generation, mothers transmit their mitochondrial DNA intact to their daughters. Occasionally, mutations may occur, and two individuals, for example, who have lived a few generations apart but share the same female ancestor, may not have identical mitochondrial DNA. This is not a bad thing though. The same may also apply to two sisters that have subtle differences in their mitochondrial DNA. We may trace the mutation in their offspring and conclude from which one of the two sisters a given individual is related. Furthermore, since the mutation rate in the mitochondrial DNA is more or less established, from the magnitude or the degree of the differences in the mitochondrial DNA sequences, we may estimate how much time, or how many generations, have passed from the time that the two sisters were alive.

Today, a certain number of mitochondrial DNA types have been identified in the living population through studies involving individuals from various continents and several ethnic groups that correspond to specific mitochondrial DNA haplotypes.[9] Analysis of the differences between these haplotypes allowed us to model the history of these mitochondrial DNAs, and thus, go back in time and trace back our female ancestors who gave us our specific type of mitochondrial DNA. As we go back in time, we can predict that the number of the corresponding haplotypes decreases progressively, until we reach a point that a single mitochondrial DNA haplotype existed in the female who gave her DNA to all living humans. This particular woman who carried this predicted haplotype, and who is apparently our great-great…great grand-mother, we believe lived in Africa about 200,000 years. She has been called by various lyrical names, such as our "Mitochondrial" or "Black" Eve. Naturally, this Eve was not the only human female who was alive at the time. She likely had sisters and stepsisters, cousins and "friends." However, she was the only one who established the mitochondrial DNA lineage that persists today, while all other females' mitochondrial DNA disappeared at a certain time of human evolutionary history. Analogous calculations have also been made for the Y chromosome that point to an Adam, and are in general agreement with the results obtained for Eve. Of course, most likely those individuals that we named Adam and Eve did not produce offspring together, and probably they didn't even live during exactly the same time. However, their DNA lineages persisted through history and it is the one that we all share today.

3.3. Archaic Humans: The Neanderthals

Based on those and various current analogous calculations, which also take into consideration the divergence of chromosomal DNA, we believe that "Adam and Eve" lived among 20,000–30,000 individuals who constituted the first population of modern humans, about 150,000–200,000 years ago. This population of modern humans, of course, was not completely isolated genetically. They exchanged DNA with other archaic humans. However, whenever they encountered populations of other archaic humans, the latter eventually regressed and the new species of modern humans predominated. Recent analyses indicate that Neanderthal DNA is actually present in the genome of modern humans, confirming that gene flow between them occurred (Green et al., 2010). What is probably more interesting is that when Neanderthal DNA was compared to the DNA from individuals that belong to various ethnic groups, it was found that it was more similar to that of Europeans and Asians than Africans. Strikingly, the Neanderthals are as close to the French as to the Chinese and Papuans, despite the fact that Neanderthals were found only in Europe. Populations from Africa still have some smaller amounts of Neanderthal DNA, at the range of about 0.5% or less, but specific alleles are enriched, suggesting first that back-migration to Africa has indeed occurred and also that the new environment selected for particular genes. In the non-African populations, it is estimated that the Neanderthal DNA ranges between 1–3%. While initially, we thought that this is mostly silent, not influencing any traits, now we know that it is associated with certain characteristics such as tobacco use and hypercoagulation (Simonti et al., 2016). An important role is also attributed to this DNA in the regulation of our immune responses, especially against viral genomes, a fact that has significance in contributing to the ability of people carrying it to adapt to new environments (Quach et al., 2016).

Quantitatively speaking, it is estimated that between 1 and 4% of the genome of Eurasians is derived from Neanderthals. Thus, gene flow between modern humans and Neanderthals occurred before the former diverged into Papuans, Chinese, and Europeans. The likely geographic location for this interbreeding and gene flow is the Middle East, where Neanderthals co-existed for some time with modern humans until their extinction, about 30,000 years ago. Yet, the continuous discovery and analysis of new Neanderthal specimens shows additional locations of potential interbreeding. Interestingly, this gene flow was not bidirectional, since it is only the DNA of the resident population, that of the Neanderthals, that was detected in the DNA of modern humans, and not the opposite (Figure 2).

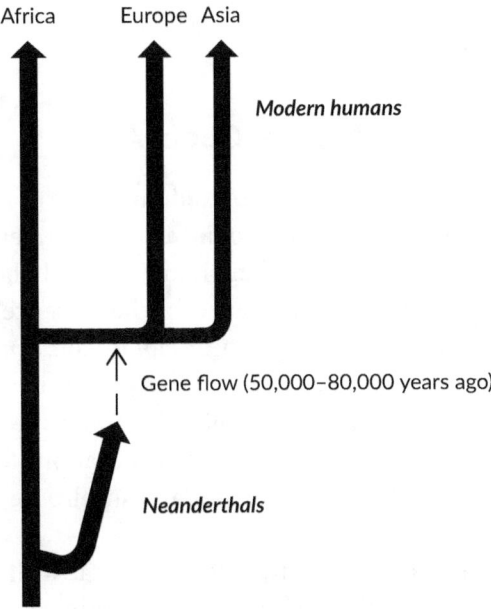

Figure 2. Gene flow between Neanderthals and modern humans.

The evolutionary ancestor of the Neanderthals is likely to be the species *Homo heidelbergensis,* which lived in Europe from 700,000 to 200,000 years ago. So, it appears that one evolutionary line in Africa resulted in the present-day humans, while the other, in Europe, to the Neanderthals. But these are all speculations that change continuously as new specimens are being discovered.

Besides Neanderthals, other species of archaic humans also existed, such as the *Homo floresiensis,* which lived in Asia (Indonesia). This species went extinct very recently, about 17,000 years ago. So, from roughly 100,000 years ago until that time, *Homo floresiensis* co-existed with humans. This species possessed a remarkably small body and brain size, as compared to modern humans, with features that are proposed to evolve due to the limited resources of Indonesia. This species, or its evolutionary ancestors, likely arrived at the island by sea in bamboo rafts 100,000 years ago (or much earlier than that, in case an ancestral species migrated), and remained there until their recent extinction.

The Denisova hominins or Denisovans is another interesting archaic human species that is highly related to Neanderthals and went extinct about only 15,000 years ago. They also interbred with modern humans, and in some present-day populations, the Denisovan DNA fraction is quite high: In Melanesians, it ranges between 3–5%, while in the DNA of Aboriginal Australians can be up to around 7%.

3.4. Modern Humans Migrate Out of Africa

Initially, modern humans were restricted to Eastern Africa, but from around 10,000 years ago, they have colonized all continents. During their history, they replaced all other archaic humans, until about 30,000 years ago, from which the most recent evidence for the presence of Neanderthals has been recorded (with the exception of *Homo floresiensis* in Indonesia, which existed until 17,000 ago, at least). All major events of human evolution have occurred in Africa, which is the geographical source of almost all genetic variation that exists today. Almost all variations that can be found in populations throughout the world today can also be seen in African populations.

If we test genetic heterogeneity outside Africa, we'll realize that it is much lower than that within Africa. An interesting observation is that genetic diversity is decreasing with increasing distance from Africa, with two major population bottlenecks having occurred that contributed to this loss of diversity (Amos & Hoffman, 2009): one around Africa and in the Middle East and another around the Bering Strait. During these bottlenecks, that are associated with sharp declines in population size, for example when groups of people migrate, genetic variation decreased considerably because only a fraction of the genetic heterogeneity passed into the new population. Thus, by analyzing genetic heterogeneity data from people around the world we came to the conclusion that the first human populations, while still in Africa, acquired a certain degree of genetic heterogeneity and then migrated out of Africa, and not the opposite, according to which heterogeneity has been generated out of Africa (Figure 3). In the resulting populations, genetic variation progressively decreased over time and distance because each time people migrated only a subset of them did so, carrying only a fraction of the diversity that was present in the "parental" group of people. Therefore, the new populations that were established in different geographic locations possessed only a subset of the available genetic polymorphisms that were originally generated in Africa.

Among the older evidence indicating the migration of modern humans into other territories outside Africa are the remains found at the Es Skhul Cave, located in today's Israel about 20 km south of Haifa. Around 100,000 years ago, other archaic humans were also present in this region, such as the

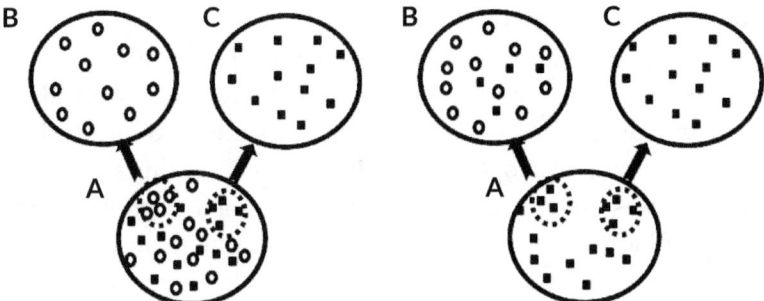

Figure 3. People in Africa have first acquired genetic heterogeneity and then moved outside Africa (left), from area A to areas B and C. Thus, both "circle" and "square" genetic characteristics of B and C populations, respectively, are represented in A. In the diagram in the right panel, the genetic characteristic "circle" that predominates in B population, appeared after the migration of people and therefore is absent from the originating population A. However, square characteristics are still there.

Neanderthals, with whom modern humans, as we said, interacted. Interestingly, between 80,000 years ago and 45,000 years ago, evidence for the presence of modern humans in the Skhul area is not available, which probably suggests that the Skhul humans retreated back to Africa from where they came from or they went extinct. Probably, the presence of Neanderthals at the time prohibited their attempts for expansion into other areas and, thus, they remained confined to Africa, which by 100,000 BC had been colonized in its entirety.

Comparisons in the anatomy of these Skhul samples and contemporary Neanderthal samples suggest that various behavioral differences had already emerged between these two hominid populations. A significant shift in human manipulative behaviors was associated with the earliest stages of evolution of the modern humans (Niewoehner, 2001).

According to the widely accepted timeline of human migrations, sometime around 80,000–65,000 years ago, humans started spreading across the world. Caveats, however, still remain, and new findings continuously cast doubt on our original understanding of when exactly these events had occurred. For example, evidence of human teeth from at least 80,000 ago has been obtained from an excavation site in Fuyan Cave in Daoxian in southern China (Liu et al., 2015). Thus, maybe dispersion of human population occurred much earlier than we originally thought, at least towards Asia. In Europe, humans entered much later, around 45,000 ago, which is likely due to a barrier imposed by the Neanderthals who were already present in the region.

Each time a migration wave occurred, a geographical area was occupied by people that had a distinct set of genetic features. Most likely, this set of features was random and reflected the genes and alleles that were present in the group of people that originally inhabited this area. Whenever a population was established in location A and then a subpopulation was migrating to location B, and subsequently to C, the final population had alleles present in the originating population, but the frequencies were often different. The latter difference is usually due to selection of specific alleles in the new environment, and is also due to the genetic drift and founder effects, since the group of people that migrated was not fully representative in terms of genetic frequencies, of the population they originated from. From this kind of data and appropriate analyses, we can calculate the route of these migratory events, the time periods involved, and the genetic relationships between populations. And while the discovery of human remains at various distal locations shows that humans have indeed migrated there, their contribution to the genetic pool of present-day people can only be inferred by the genetic analyses. In other words, being somewhere does not necessarily mean that these humans established populations with a continuum until today.

A major migration wave around that time occurred via the Horn of Africa and involved the dispersal of humans to the Arabian Peninsula, towards coastal areas of Asia. Until around 65,000 years ago, they had successfully colonized Australia. According to an interesting, finding from a DNA analysis of hair of an Aboriginal man from Western Australia, the colonization of Australia occurred quite early and took place even prior to the separation of European and Asian populations (Rasmussen et al., 2011).

The similarity between Indians (from India) and Europeans, as regards their "Caucasoid" features, traces its origins to the similarities of the Indians with the groups of people that originally left Africa during this migration event. The reasons why the humans at this time preferred this route for the exodus from Africa, as compared to the alternative and straighter route that could involve spreading through the Middle East, are unclear. Probably the presence of Neanderthals in this area represented a definite obstacle for their trip. In addition, geographical reasons may also account for this preference. For example, the sea levels of the Red Sea were considerably lower than those today, which might have facilitated their travel through that route. Genetic evidence, however, is clear. Mitochondrial DNA analyses implies that the haplotypes of human populations in these areas that have been colonized during the first colonization events and involved southern Arabia, India, and Asia are not present in populations from the Middle East. Naturally, these haplotypes are present in African populations. If the route of this migration was through the Middle East, we would have expected to see their genetic fingerprints in the Middle Easterners. Thus, we conclude that the Middle East was not part of their travel at that time.

While at first the modern humans were concentrated in the coastal areas of Asia, eventually they started spreading north, and by about 40,000 years ago, they occupied the interior of Southeastern Asia, while by about 25,000 years ago, they reached and colonized areas around modern-day Beijing. Of course, the genetic composition of the people that occupied Asia was far from being considered stable over time. Technological progresses and, importantly, the establishment of agriculture between 7,000 and 6,000 years ago around Yellow and Yangtze Rivers, led to the rapid expansion of Northern Chinese populations and their intermixing with adjacently located populations. Vice versa, the repeated conquest of China by tribes from the North, such as the Huns and the Mongols, also contributed to the increased genetic complexity of the people that occupied Asia.

The colonization of Japan is likely to have occurred through two different waves of migration. The first wave, around 30,000 years ago, was likely from people north of Australia and Southeastern Asia, who crossed into Japan by a land bridge that existed until 12,000 years ago. These people, who are called Jomonese, were cut off from the Asian mainland for more than 10,000 years and were mostly hunters and gatherers. Around 2,300 years ago, another group of people called the Yayoi entered Japan from the Korean peninsula, bringing with them rice, agriculture, and metal tools. These two major populations that occupied Japan, the Jomonese and the Yayoi, eventually intermixed, and the latter population predominated over the former. It is estimated that Japanese people today carry up 90% Yayoi and 10% Jomonese DNA.

Modern humans re-entered the Middle East from Northeast Africa around 45,000 years ago. Neanderthals who were present there at that time progressively disappeared, and modern humans were the only occupants of the area. Around 40,000 years ago, they reached and colonized Europe from the Middle East, either via Greece and the Balkans or via the Black Sea basin and the Ukrainian plains. This is supported by the fact that mitochondrial haplotypes in Europeans are mostly the derivatives of the mitochondrial haplotypes from individuals from the Middle East.

The colonization of Europe has an interesting twist. The first migration wave towards Europe, which eventually resulted in the progressive disappearance of the Neanderthals between 40,000 and 30,000 years ago, consisted of modern humans that culturally, although more advanced than the other archaic humans, were still hunters and gatherers, thus relying on the same resources for their survival as the Neanderthals. While at this time modern humans occupied virtually all Europe, the Ice Age that was in its peak around 20,000 and 16,000 years ago pushed people towards the warmer areas of Europe, close to the Mediterranean and north of the Black Sea. This retreat into the southern territories lasted until around 13,000 years ago, when the Ice Age was gone and people, assisted by the gradual temperature increases, the melting of the glaciers, and the expansion of forests, moved back again into Northern Europe. Those people are, at least in part, ancestors of modern Europeans.

The other major part of the genetic pool of modern Europeans traces its roots from another major migration wave, originating again from the south and likely through the Middle East. Around 10,000 years ago, a major revolution took place in the Middle East involving the invention of agriculture. Interestingly, between 10,000 and 4,000 years ago, the development of agriculture had occurred in many different societies all over the world, from the Middle East to China, to Africa and to the American continent. This development had dramatic consequences on the lives of the people and the progress of civilization. Larger cities were built, societies demanded a more sophisticated mode of organization, and people were producing more than what they needed in order to survive. So they had time for other activities, as well. In addition, a considerable growth in population occurred that caused the need for geographical expansion.

It is likely that, following the invention of agriculture, whole populations from the Middle East were already practicing it, as opposed to single individuals who could have transmitted this knowledge, moving north and then west into Europe, in coastal areas or areas close to rivers at which people could establish societies that were based on agriculture. As early as 9,000 years ago, according to the archeological evidence, farming was practiced efficiently around Athens and progressively spread to other coastal areas west of Greece. It is likely that the spread of agriculture was not due to the immigration of individual people who knew how to practice it and that taught the farming practices to the locals. It is rather likely that the migration of populations of farmers from the Middle East towards Europe took place at this time. Those people intermixed with local people whenever they encountered them, absorbing them. The genetic evidence is consistent with this hypothesis, since it shows that a genetic gradient is present in Europe from south to north, with certain alleles predominating in Southern Europe, in areas that can support farming, that progressively give way to alleles that predominate in Northern Europe, where hunting and gathering represented the common practices. It is quite likely that the present heterogeneity within the various European populations, which ranges from physical characteristics and extends to certain mentalities, is due at least in part to the genetic origins of these populations and their relations to the aforementioned migratory events.

This hypothesis regarding the genetic heterogeneity of the populations of Europe was originally proposed by Luca Cavalli-Sforza, a pioneer human geneticist and biological anthropologist, and caused many controversies until its acceptance (Cavalli-Sforza, 2001; Cavalli-Sforza & Cavalli-Sforza, 1996). Today, according to the results of genetic studies at which Cavalli-Sforza's lab played the major role and involved studying the variation in Y chromosome of European individuals, we believe that following the original colonization of Europe, which accounts today for about 10% of the genetic pool, the major genetic constituent originates from the continuous expansions and retreats of people from the Middle East during the Ice Age. The final migration event, during the agricultural period, represents about 20% of the variation of Y chromosome and mitochondrial diversity and is due to the massive arrival of the farmers from the Middle East. Of course, these subpopulations of Europeans were never isolated, but continuous and extensive intermixing always has occurred. Thus, it is not possible to identify individuals that trace their origins exclusively to either the farmers or the hunters of the pre-agricultural period. Even in the Basque population, which is considered the direct descendants of prehistoric modern humans, alleles from the Middle Eastern farmers can be traced.

From that time until today, people from other continents are continuously entering Europe, leaving their genetic fingerprints. The Huns, for example, were nomadic people of Asian origin who invaded Europe around the fourth century, reaching France and Northern Greece. The origin of Huns in Asia is vague, but it is thought that they can be traced to the Xiongnu people in Western China, who had been defeated by the Han dynasty in the first century before moving towards West. Today, several people and ethnic groups claim their association with the Huns, such as the Hungarians, who in their national anthem describe themselves as "blood of Bendegúz," the father of Attila, the Huns' glorious king. The Turks are also thought to be related to the Xiongnu people, starting their migration from areas of North China and Mongolia towards present-day Turkey. This migration was initiated around the sixth century and lasted until around the eleventh century. Today's Turkish people can trace their origins to those Turks, but also to people occupying these areas at the time of their invasion, such as people of Greek and later Byzantine descent.

Migration into Europe during the latter historical periods was not only due to forced invasions and eventual interbreeding with the local populations, but also the result of immigration in order to satisfy economical and other demands of the people involved. France, for example, especially during the twentieth century, hosted several immigrants, not only from other European countries but also from Northern Africa, at a period at which the

Industrial Revolution generated the need of increased manpower to work in the developing industry. Even when these purely economic needs ceased to occur, the immigrants stayed, since they had adopted the life and culture of France and the other countries that received them. A similar wave of immigration is currently underway in Europe from Asia and Africa, which is empowered not only by economical but also by other political demands, as well as from the Far East, such as from China. The latter usually do not arrive in Europe as typical labor power, but rather as traders or skilled workers in scientific and technological disciplines, and it is anticipated they will eventually assimilate into the local societies and populations.

The migration into the Americas started around 20,000 ago from people originating from Eastern and Central Asia. They entered the American continent from the north and then started moving south until they established populations on the whole continent. It has also been proposed that European people may have crossed the Atlantic, reaching South America first before moving north. Settlements of Vikings may have also been established around the tenth century in North America, but it is rather unlikely that they endured as permanent colonies, leaving a considerable genetic fingerprint back.

Those migration waves established the Native American populations that occupied the Americas until the sixteenth century and gave rise to significant civilizations, such as the Mayas and the Aztecs in modern-day Mexico and the Incas in Peru. The (re-)discovery of America by Columbus initiated a second, massive wave of colonization by Europeans and Africans (as their slaves, initially) that continues today. Immigration from all over the world, including Asia, during the twentieth century also continues into the present day creating a genetic diversity pool of increasing size in North America.

The European colonization of South America resulted from Spanish and Portuguese migrants, who brought with them more slaves from Africa than their Northern and Western European counterparts did who settled in North America. Interbreeding between the European and African people in America was originally much stronger in Central and South America than it had been in North America. Following the European colonization, local populations in America were either eradicated or absorbed, with only a few and limited subpopulations—or, even more precisely, groups of people—retaining their ancestral identity culturally and, to an even smaller degree, genetically.

Collectively, with these processes, the whole globe has been colonized by groups of people that eventually became specific and relatively well-defined populations. These populations carried alleles at frequencies that reflected in their rates, and occasionally in their identities and composition, the alleles of the original populations that at the very first instance moved out of Africa. Through the progressive change in the allelic frequencies, we are now able to monitor and recreate these events and extract conclusions regarding their timing and their magnitude.

None of these populations ever lived in complete isolation for prolonged periods, as supported by the cultural and, probably even more importantly, the genetic findings. A continuous flow of genes always occurred and is still ongoing at variable rates and directions between the various populations. Occasionally, this is related to certain socio-economical conditions that affected (and reflected to) human history. Furthermore, the frequency of these alleles in the various human populations changed and continues to evolve over time by phenomena involving genetic drift and selective pressure. The variation of human populations, as we witness it today, is due to the set of alleles that predominate each of these populations, and ranges from genes affecting physical characteristics and likely behavioral tendencies as well. In turn, the latter affect the development of specific human cultures and societal norms.

Therefore, the current distribution of people around the world is the cumulative result of all these great, and occasionally massive, migrations we have described. Always, the common denominator is that a group of people of a variable size that is a subset (fraction) of the originating population was moving into another territory. In general, if the size of the population of the people that migrated is small or the selective pressure in the new environment is strong, then the divergence of the new population from the originating one is high.

On their way, these people frequently interacted genetically with other people of the local populations. This interaction, in the form of minor or major genetic exchanges, also occurred at a varying degree in the final destination as well, in the territory at which they migrated and settled. In all cases, however, the people that have migrated were not representative genetically of the originating people. It was only a subset of the genetic traits that were carried with the people and, in general, the smaller the group, the less representative it was of the originating population. Otherwise, we would have expected all non-Africans to be more like Africans, and since at some extent we are all Africans, we would expect all non-Africans to have more similarities among them. In other words, non-African populations should have been more uniform, while differences between at least the non-African populations should have been kept at a minimum. What actually has happened, though, is that these founder people who relocated into the new territory carried with them specific alleles. If, let's say, their ratio of a given allele was 1:10,000 in the originating population, but during a migration happened to be represented even once in the 1,000 people that migrated, then its ratio instantly increased by a magnitude of 10 (1:1,000 vs. 1:10,000). In that case, this allele is enriched in the migrating subpopulation, and the trait associated with it is more common in the migrating than the originating population. And this enrichment happened without any selective pressure, meaning that by having it or not in the new territory, no difference was made for the individuals bearing it. Of course, as mentioned before, the frequency of this allele is not going to remain stable over time and generations, but it will change eventually, depending on the selection pressures in the new environment, the genetic frequencies of the corresponding alleles in the "receiving" population, as well as the genetic drift. The opposite case is also quite likely, in which an allele is present in the parental population at a low frequency, but it is not represented anymore in the migrated people.

We emphasize again the fact that all these notions should not be viewed as treating African or other originating populations as homogenous. This is an idea that is distal from reality, since in Africa several distinct subpopulations can be identified with frequently very different characteristics. Think about the Pygmy people in Africa with an average height of about 1.50m and the Dinka people that have a height of about 1.80m. Therefore, the divergence between the migrating and the originating populations increases even more, depending on the specific African subpopulation that has initiated the corresponding migration wave.

These phenomena, of course, occur not only with the alleles that are related to physical traits, but also with those linked to diseases, as well as to any other characteristics, including the behavioral ones, that may have a genetic component or constituent. Sometimes, these specific characteristics are associated with an advantage that can be more pronounced in the new location, and thus, to increase even more in frequency over time. Alternatively, they may be related to the acquisition of a disadvantage, and thus exhibit a tendency for stabilization or reduction in their frequency.

Take again the skin color as an example. The alleles that control how dark the skin color is are already there, in the originating population in Africa. An evidence for this is the high variability in the skin color intensity today in different African people. It is easy, however, to imagine that many thousand years ago, when people decided to migrate out of Africa, life would have been more comfortable for those colonizing Northern territories to have less-dark skin color, as compared to those with darker skin color and living towards the north. Not that life would be intolerable for the others. Thus, today, we can still see people occupying areas with similar distances from the equator and having considerable differences in skin color.

In a similar manner, if a gene or combination of genes were associated with "milder" behavior that was less prone to risk-taking and exploratory activities, we would rather see these people being more comfortable with agricultural activities, as compared to hunting and gathering food, which represent activities with higher risk. This increased risk may refer to both the uncertainty in the satisfaction of the nutritional demands and to the fact that occasionally it may subject individuals to life-threatening or dangerous conditions. And, as regards agriculture vs. hunting and gathering, things get even more complicated if we take into consideration additional phenomena and patterns that must emerge, such as the collectivistic behavior that appears to be an important adjuvant for the development and sustainability of the agricultural societies and the fact that complex governance systems must develop and followed by the corresponding people. Obviously, for the operation of activities such as farming, the intuition is not sufficient. It has to be accompanied by some empirical knowledge of how to do it, being in an area at which the climate and the soil permit agriculture and, naturally, not being in the critical proximity with neighbors that sooner or later will force the relocation. That some of us prefer fishing and hunting (yes, they both frequently go together!) over gardening and farming as hobbies may reflect, to some extent, our genetic constitution. While hunting is related to immense action, short-term planning, and intuition, farming is associated with patience and long-term planning. While the first is mostly an individual's activity, the latter is usually collective and is highly dependent on the context.

CHAPTER 4

The Rise of Personal Genomics

Depending on the specific scientific discipline according to which people attempt to interpret large-scale choices and decisions, some may recognize the climate, the geophysical characteristics, or even the presence of extraordinary persons within specific populations as the primary reasons that causatively shaped those decisions. Even coincidence may have played its role in the outcome of certain events and historical fates. Imagine, for example, how different the present world would have been if Alexander the Great didn't die at the age of 33.[10] Or if a fire in 1901 did not destroy all but one prototype in Oldsmobile's manufacturing facility, sparing only the one that resembled the relatively cheap cars that targeted the average American and which shaped Oldsmobile's future commercial directions.[11] In the first case, we may had seen Greek as the most common language in the present world, while in the second case, we may had considered Oldsmobile as a premium manufacturer of luxury cars, leaving more space to other popular automobile makers.

Among all factors that shaped history, biology is the one that receives the lesser attention. Yet, the presence of some genetic predisposition also played its role in the fate and outcome of history, particularly in the cases that human choices—direct or indirect and active or passive—were made. And this effect, even if it is seen as minimal when viewed during a certain chronological instance or when evaluated and assessed in a given snapshot of human history, its consequences were magnified and multiplied at extended historical periods and studied in the context of large human populations.

The premise for such notion is that the various different groups of people, from the time they reached a new location and onwards, are starting their specific history by already having different tendencies, disease characteristics, and other traits that may be slightly distinct from the population they came from. More importantly they are distinct from other groups that have also migrated at a different time. In other words, they are unique to some extent at their capacities, not just as independent individuals or the sum of a certain number of individuals, but rather as a whole that constitutes a dynamic group and entity. And these differences become even more pronounced when we look and compare different populations that, although they came from the same pool or originating population, they correspond to different major migration events. So, the genes that migrated with them although qualitatively were similar, their frequencies were different. This applies more or less to all different populations in the world but is more prominent between the Asians and the Europeans, or Easterners and Westerners that both correspond to populations that originated from Africa (like all populations in the worlds) but chronologically are distal by as much as 30,000–40,000 years, were geographically isolated quite well and apparently, they were at such numbers that could establish cultures and civilizations pretty independent from each other. (Figure 4).

Although they were coming from the same area, they carried a different set of genes with them that defined their whole existence. It could have been that this difference is because the founding (migrating) people were genetically different in these two cases. Maybe, different selective pressures were also applied during their travels and settlement that shaped the final

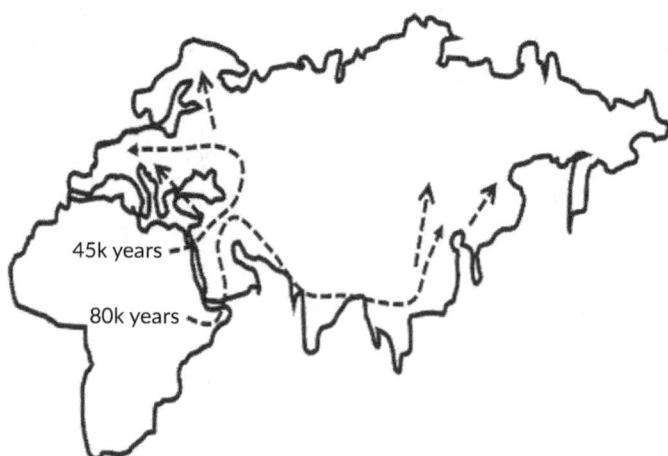

Figure 4. Around 80,000 years ago, people from the Horn of Africa migrated towards the Arabian Peninsula and coastal Asia. Around 45,000 years ago they colonized Europe.

populations differently, or even more likely, a combination of both. Nevertheless, the difference is mostly quantitative and not qualitative. In different groups, including "Easterners" and "Westerners," it is not that specific genes (or alleles to be more precise) are present in one and absent from the other but instead that their overall frequencies are different in the two populations. As we mentioned already, the genetic variation between different regions is smaller as compared to intrapopulation variation. This makes things even more interesting because it highlights how significant a "dosage effect" is, and implies that different traits should be in some kind of balance with each other in order to be able and function efficiently. This "balance" ultimately determines how functional and efficient these populations will be in terms not only of their strict biological characteristics but also to the whole cultures and civilizations they will create.

Notwithstanding the details of this continuous—and actually never-ending process, as we can evidence it today—the fact of the matter is that in different populations, that largely correspond to different geographic locations, the various genetic alleles exhibit different frequencies. If we project this at the level of a single individual, we can extract conclusions regarding the genetic ancestors of this person and his or her ethnic identity. The latter does not work in the sense of attributing that person a single ethnic origin, but rather by determining his similarities and differences from various different ethnic populations and based on that by offering a genetic similarity result with the people that live in corresponding geographical locations. It is quite interesting how this works for people in the U.S. that can find out their roots in the country from which they or their ancestors immigrated from.

Several genetic analysis companies operate today that provide this information. To use their service, one can submit a DNA sample, in the form of a saliva wash, and get results on the composition of his genetic markers and how common or uncommon they are in the various ethnic groups for which such scientific data are publicly available. Specifically, one can get a report at which the percentile of his African, Asian, or European DNA composition has been calculated. As more and more genetic data become available one can receive even more detailed information regarding their similarities with people in the Eastern part of the Aegean Sea or the northern side of Greece, thus tracing their roots with high precision. With this information, in association with global data on DNA allelic frequencies that are available, he can get an idea regarding the specific populations that are more similar to him,[12] or with which ethnic populations a better "match" exists. Not surprisingly, by such analyses we—as Westerners—identify Eastern polymorphisms in our genome and vice versa. What we expect, though, is that people of European ancestry will have collectively, polymorphisms at frequencies that are more "Western"-like in their genome. This is an undoubted proof that our genetic fingerprint or signature offers evidence regarding our genetic history, our ancestors, and eventually our roots and origin.

As we have mentioned explicitly at the beginning of this book, the commercial "value" of studying and analyzing genetic polymorphisms is not only limited to their inherent property as genetic markers that is useful in order to trace gene flow and to identify genetic ancestors. Of particular significance is also their use as the actual determinants of variation when their variability is translated to different properties. Numerous genetic studies attempt to link diversity in the DNA with variability in physical characteristics, predisposition to diseases, and behavioral patterns. Based on these types of data, another service that the DNA-testing companies can offer is that they calculate the probability of getting sick from specific diseases. In order to do this, they identify the presence of specific markers in an individual's DNA. It has been established by several genetic studies that specific alleles in certain genetic loci are more common in patients that suffer from a particular condition. Thus, by carrying this allele, a person is more (or less) likely to get this disease as compared to the corresponding probability for a person who doesn't have it. The idea behind such service at the first instance is to practice the "know thyself" notion and satisfy the individual's curiosity at a sub-cellular level on his or her genetic makeup. Of even more practical value is that by knowing the most likely causes of sickness, one may be protected better by adapting accordingly his or her lifestyle and avoiding certain high-risk habits. For example, avoiding "junk" food is good for everybody, but even more imperative for someone at increased risk for atherosclerosis. The spectrum of diseases that are included in such analyses continuously increases, and ranges from heart diseases to some cases of glaucoma and are extended to various cancers.

Another type of information someone can get from this analysis is related to pharmacogenomics, or to the prediction of the response against specific drugs, such as beta blockers, warfarin, and statins. This is a type of information that is of particular value when clinicians are determining the doses of the drugs to be applied or identify effectiveness and undesirable effects.

The precision and the range of conditions that these DNA analyses will be able to predict are only expected to increase in the future. Naturally, these DNA-based approaches occasionally touch sensitive issues and warrant increased caution in their use, particularly when they attempt to answer questions extended beyond simple curiosity or, alternatively, of a clearly defined clinical value. Potential categorization of people with respect to their DNA profile is an imminent danger that reminds us of the shameful attempts associated in the past with eugenics.

Related to these ethical issues or based on purely scientific and technical grounds are the setbacks in the genomics industry. In the summer of 2011, the UK Border Agency canceled the use of DNA testing for purposes related to the identification of the country of origin in individuals requesting asylum.[13] In that case, the home country (or ethnic origin) of an individual could be pointed by the frequency of certain genetic markers. An analogous setback in the DNA testing industry occurred in 2010 in the U.S.A., when attempts for direct-to-consumer DNA tests had been postponed.[14]

The problem here is not the actual result of the test, in terms of the identification of a specific allele, but rather the interpretation of the findings and whether they convey medical or a general wellness information. The main arguments against the validity and the specific value of the information of the results of these tests are related to the fact that most of the diseases and traits, especially the behavioral ones, reflect the complex and collective output of many parameters at which the genetic contribution represents only a single subordinate. This is definitely true and is reminiscent of classic and ongoing debates on the "nature or nurture" contribution for the definition and shaping of various traits, including that of human personality.

For the present purposes, it is not that important to take sides on this debate and claim whether nature (the genes) or nurture (the environment, physical and social) is more important. But even by accepting that genes indeed play some role, however minimal, in the building of our behavior at the level of individuals, I think that this is sufficient for making a considerable contribution in the outcome of certain historical decisions, if viewed at the level of populations that impact through time whole cultures.

In order to appreciate the actual contribution of the genes in the behavioral patterns of populations, we always have to take into consideration two important parameters:

a. The *time* we refer to is historical time and, thus, it is by far larger than that of the time period of an individual person's lifespan. To that end, we need to consider that the choices made by a given group of people and at a certain time may affect profoundly the lives of the subsequent generations. In turn, individuals of subsequent generations, by making analogous choices, may establish a trend, or a norm that generates a behavioral and cultural pattern. In other words, there is a consistency in the choices. Take agriculture as an example that, as mentioned before and will also be described subsequently, is associated with the onset of collectivistic societies that likely posses a behavioral (and likely genetic) subordinate. When at the very beginning of farming some people benefitted from practicing it, the subsequent generations had eventually to "choose" from continuing it or to seek for alternatives. Those that were feeling comfortable being farmers could have continued from the point that this (as an advancement) was left from their ancestors. They did not have to re-invent agriculture from scratch. So, they could benefit even more from that, especially those who were more efficient in practicing it. And since, at a certain degree, it was presumably in their genes to feel comfortable with farming and living in the associated types of societies, a trend had already been established. The latter might also produce a "selective pressure" consistent to which the individuals that could operate better in such societies might also be more efficient in advancing them. Thus, in the long run, these societies could not help but become agricultural societies. A similar tendency may be applicable to various types of societal organization that are associated to the development of distinct cultural norms. The individuals, for example, with a tendency to nomadic life are more efficient in nomadic societies than those with a tendency for being settlers. Of course, those that might feel better as settlers, by taking advantage of the presumed lower competition from other settlers, could have established settlements quite efficiently. In that case, though, we may anticipate that a subpopulation has split from the original population. And this subpopulation might also possess slightly different frequencies in the corresponding alleles that are associated with this behavior. The complexity of these processes increases by considering additional factors as well. For example, societies practicing extensively and relying highly on agriculture require certain types of governance that are linked to the establishment of cities of higher population, the advancement of trade, and other characteristics associated with this. All these, are practiced more or less efficiently, depending among others on various personality traits that make individuals suitable for those. Thus, for their efficient implementation they need to be accompanied by a series of characteristics that have some complementarity with each other. And those societies that combine them better are likely more efficient and enduring. In addition to that, another consideration is that efficient societies require diversity. They need savers and consumers, leaders and followers, worriers and warriors. Thus, the factor of frequency enters that modulates the outcomes.

b. The *number of people* we refer to, that is, extended to the size of ethnic populations. Within that view, we anticipate the effects to be cumulative. In order to understand this, we may consider that a trend followed by many people, in large populations, is harder to change than that followed by only few, in smaller populations. Therefore, it is not only the actual proportion that matters but also the basis of people who follow it. This may also have to do with other factors, but relatively speaking, think about the following example: During the COVID-19 pandemic wearing masks became recommended and imperative. Adherence to this, though, was followed more in larger than smaller communities. Responsible for this can be the relative resilience of smaller populations to follow guidelines, the false sense of security they feel, or the fear of exaggeration in larger cities at which anonymity prevails. These can also be factors affecting the transmission of behavioral traits in larger populations. Thus, cumulatively, the consequences are larger than we may originally think.

Later, we will come again to these points when we discuss the individual's traits and how they can possibly define a population's traits.

CHAPTER 5

Greeks versus Chinese: The Prototypic Behaviors

The fundamental idea of this book is based on the premise that two different populations, which differ in their ethnic origins and thus in their genetic "fingerprint" or identity, are likely to respond differently against the same or similar external stimuli. These differences are likely to be interpreted and reflected as different decisions through the course of their history, which practically means that even under identical conditions they will make different historical choices. Such choices are made frequently by the leaders of the corresponding societies, but they indeed need to have the support of the people and to eventually reflect the individuals' mentalities.

Naturally, such arguments suffer from various conceptual and methodological limitations, and they are quite vulnerable to rigorous academic criticism. For example, the statement "under identical conditions" reflects a rather experimental approach that, in real life, is practically and theoretically impossible. Thus, by experimental means such an argument cannot be proved or disproved. It's meaningless to argue that we may put the same group of people under exactly the same historical context and monitor their reactions. Even if this were possible, the linear course of history would have incorporated the result of the first reaction as a historical experience that unavoidably would have affected the outcome of the—chronologically—second (latter) reaction. However, with such an argument, provided all specific and general limitations have been taken into consideration, we may identify some trends. For example, the Huns can be considered as a population prone to fighting, confrontation, and war. They definitely had their impact in history, but this was essentially related to their military expeditions. Undoubtedly, the geographical and cultural context in which this tribe or group of people developed historically played its role in shaping their tendency toward warfare. It was essentially a matter of "natural selection" for them in order to continue existing. But again, by accepting this we introduce biology as a variable. We may anticipate that, provided this behavior is at least in part due to their genes, even if they had to live in an isolated community, such as a "peaceful" island (let's say in the Pacific), again they would have exhibited signs of this warrior behavior. In a similar manner, we usually view Chinese as people that exhibit a tendency for stability as a society, with a certain dislike against revolutions that cause drastic and rapid changes in the social structure. This may explain why Chinese society and political structure had this continuity over the years.

At the same time, we understand ancient Greeks as people who enjoyed argument and critique while continuously attempting to pursue their personal and individual autonomy. This was highly reflected in the essence of their gods that "suffered" from human emotions and that they were taking sides at war suggesting that there was not a single "moral" outcome in their conflicts and confrontations. Even their philosophers, that all had their several followers, while they lived during the same period and many of them at the same time, they expressed quite opposite ideas implying that there was not a single idea that could represent the society's mainstream thinking. Their political system as well reflected this fluidity as well. They are known, of course, for the invention of Democracy. However, they also had a monarchy, an aristocracy, and even a *coup d' etat*. More importantly each one of them had its own moral and ideological supporters. By being the products of our current society, today we have acquired tendencies to support one system or political ideology over another, but this was not the case in ancient Athens in which such biases were not intense. These differences are condensed to that while for Confucius, the main advice is obedience to elders and social superiors, for Socrates (teacher of Plato who was teacher of Aristotle) is the acknowledgment, that the only thing he knows is that he knows nothing!

Our hypothesis is that these differences are not random and stochastic but are dictated—or if this is too strong an argument, they are influenced considerably—by the genes that these people carry. Therefore, what we view as a choice is rather the complex and collective outcome of the influence of the peoples' specific genes, combined with the effects of their specific environment. In turn, this makes the probability for rendering a certain choice distinct between different populations.

Let's assume that there are two hypothetical populations, named A and B, that can make a decision to take an exploratory action K or not to take this action, being the decision L. If there is actually a gene that increases the tendency to pursue the exploratory activity, and this is more frequently found in the population A, it is anticipated that these people will be more likely to take exploratory actions, while as a population to be committed to exploratory activities more frequently. And this tendency will not be expressed only once but repeatedly. Since this feature is "programmed" in the genes, then their descendants (as well as their ancestors) will be more likely to take this exploratory action than the B people. Thus, it is conceivable that A people will have a history that is denser in exploratory events than the B people. It seems that a certain pattern emerges in that case that will profoundly affect the culture and civilization of the A people. By analogy, imagine a population with a higher frequency in a gene that offers a tendency for a "collectivistic" behavior. Such behavior implies that the individual is defined by its position in a certain group, and frequently the needs of this group surpass the needs of the individual. On the opposite side is the individualistic behavioral pattern, at which the focus on personal autonomy plays a major role in one's life. In that case, it is quite likely that the population with such a collectivistic gene in abundance will display a tendency toward developing a culture at which individualism is reduced or diminished. This, in turn, will cause the development of less liquid societal structures at which people maintain relatively strict positions. Furthermore, the good of the group will be predominant while the good of an individual will play only a minor role. It is also quite likely that such societies will be larger and characterized by a type of more centralized administration that will maintain its structure more efficiently. On the other hand, the populations that have the "individualistic" gene at a higher frequency are expected to be driven by the pursuit of personal autonomy; relationships between individuals will not be stable and constant but rather transient. Furthermore, in such societies the individual's needs appear to play the major role, regardless of whether they are good or benefit the whole group.

In general, these differences do not simply affect choices such as population expansion against maintenance, attack or defense, or development of complex and hierarchical societies, but can trace their differences in the very essential notions of how the world is made and operates, what are the essential driving forces of the universe, what is the purpose of life, as well as to questions related to the pursuit of happiness and how this can be achieved.

As we will see in subsequent chapters, such polymorphic genes with different frequencies in various populations actually do exist. For example there is a specific gene that is associated with elevated tendency for sensation-seeking and risk-taking. The fact that this gene is polymorphic means that certain versions of this gene are more likely to be associated with increased risk-taking than others. Depending on the exact context of the experiment that led to such conclusions, this particular gene has been linked to diverse actions ranging from the degree of pleasure that someone receives during specific thrill activities, to gambling and even drug abuse. Thus, it is quite reasonable to speculate that such genes will have a profound and pleiotropic effect capable of affecting diverse facets of life. To that end, a whole spectrum of behavioral patterns will be expected to go together in the corresponding individuals, a fact that makes the differentiation of these people even more pronounced. More importantly, such genes in order to operate efficiently in the society, they need to be accompanied by other genes that offer complementary characteristics.

Before we try to view and probably attempt to understand and explain different populations and cultures in terms of their different genetic identities, it is important to identify such elementary notions that are distinct among populations. In other words, identify differences in the essence of distinct populations and realize that certain analogies are rather superficial and oversimplistic, despite our tendency as Westerners to view and interpret history and people by our rather naïve Western eyes. For example, the Olympic games in ancient Greece had a completely different meaning than the riding or hunting competitions of the Mongols, while the geishas in the Far East possessed quite a distinct role in Japanese society, as compared to the role of prostitutes in the society of ancient Roman, despite that, due to oversimplifications, certain analogies can be identified, occasionally exaggerated and underscored.

Such notions are expected to transcend everyday lives and also to provide the context, and the mainframe at which the whole civilization, culture, and historical decisions have been made. Then, it will be probably easier to interpret different reactions in terms of the fundamental notions of how life is viewed. In order to do this, we may identify basic differences in certain civilizations and then attempt to evaluate whether these differences can be explained by the presence of different behavioral traits, due to distinct allelic frequencies, among these populations.

Among the various civilizations that emerged throughout human history, two major lines of thought can be identified that differ considerably in the fundamental perception of how the world was viewed and explained: the Eastern and the Western view that trace back their origins in the philosophical traditions of ancient Greeks and Chinese. In turn, they are exemplified by the lines of thought condensed and initiated by Aristotle and Confucius, respectively. Throughout history, these lines of thought dominated the development of these civilizations and continue to affect and shape the lives and perceptions of modern peoples as well. Not coincidentally, these major cultures were developed by populations with relatively distinct genetic identities and allelic frequencies at various loci, which had minimal interactions until only very recently, perhaps a few hundred years ago. Thus, they had the chance to develop culturally by maintaining their "independence" to a certain degree.

Contemporary globalization has scaled down these differences, with Western thought appearing to play a more dominant role today. However, Eastern perceptions of the world, although likely better hidden as such, also seem to affect Western thought and appear attractive and appealing to the Westerners. For example, small feng shui gardens containing running water and bonsai trees frequently decorate the headquarters of multinational corporations, while McDonald's restaurants flourish in Shanghai and other major cities in the East. Cross-cultural exchanges are not limited to lifestyle trends but transcend deepest notions of our civilization with important consequences in shaping future cultures and societies. For example, Eastern-type holistic perceptions gain more and more ground in Western science and find application in medicine, ecology, and even biology. On the other hand, in the East, analytical basic science research institutes, which did not represent the stronghold of Eastern science until recently, are flourishing in the last decades in Japan, representing the front of a wave that soon will involve China, Korea, and other Eastern countries.

Whether these two great individuals, Aristotle and Confucius, are the ones who indeed shaped the corresponding thoughts and philosophies, or if it is the specific societies that allowed them to flourish and express their ideas, thus dominating the corresponding cultures, is under debate and represents a deep philosophical question that is initiated by our perception on how the world and history is moving forward. My personal opinion favors the latter notion, that it is not Aristotle per se but rather Aristotle combined with the specific receptive population that resulted in the specific outcome we witnessed and recorded. It is likely that this represents an Eastern-like, holistic view according to which it is not the individual but rather the combination of the individual with the context of his society or group that plays the signifying role. But if the exact same Aristotle was born in ancient China, he wouldn't have been the Aristotle with the recorded magnitude of contribution to our civilization. The mere presence of Aristotle in ancient Peking would not have been sufficient to turn the Chinese, as followers of logic, leaning toward individualism and autonomy. The same applies regarding Confucius. His presence in ancient Athens would likely have been insufficient to turn ancient Greeks into collectivists, applying holistic perceptions of life and thought and being seekers of harmony and balance.

As an argument, I would propose that in ancient Greece, individual philosophers with more similarities to Eastern philosophies, notions, and perceptions occasionally appeared, such as Heraclitus, who introduced the concept of constant change. This notion, in its essence, is rather Eastern, since it implies that nothing is absolute and independent of its context. Everything, according to Heraclitus, acquires its true meaning only within the frame in which it is examined, while two identical conditions cannot formally exist, since at least the context will be different.[15] Despite the teachings of Heraclitus, the line of thought that was exemplified by the Aristotelian teachings is the one that actually persisted until today, and we, being Westerners, consider ourselves his intellectual descendants and the Western civilization a derivative of Aristotelic Logic.

I couldn't have explained this adequately by the more attractive and influential personality of Aristotle, but rather by the presence of one more susceptible to these ideas and minds of ancient Athenians. Vice versa, in ancient China, Ming-jia and the Mohists advocated the pursuit of truth and logic for the sake of the individual, an approach that is rather Western-like; however, their influence was not as profound as that of Confucius in their societies (Lai, 2008).

Richard Nisbett's stimulating *The Geography of Thought* (2003) very nicely illustrates the differences in the way of thinking between Chinese and Westerners, while provides a rather psychological explanation for these differences. While the essence of Greek philosophy is to understand the fundamental meaning of the world by logic and subsequently abstraction, in Chinese philosophy of the pursuit of harmony predominates against truth. Not that a "false" argument is preferred over a true one. It's just not that important whether something is true or false. What really matters is whether it is good for the group and promotes harmonious living.

The importance of harmony is detected in various different aspects of life, not simply the everyday life, but also the myths and the legends of each culture. Take as an example how dragons are seen in the East and the West. In the East, they are considered to be wise creatures with mythical powers, are valued for their magic abilities and their beauty, and are perceived without having the inherent tendency to inflict damage or to represent evil. Contrary to that notion, in the West, dragons are fearful, evil creatures, they are particularly powerful, and occasionally they fulfil punishments, or they function as guardians of a treasure or of a sacred entity or object. That difference is not irrelevant to how different cultures see the world. In the East, within the context of the world's harmony, people do not have to be scared of the dragons, which can co-exist harmoniously with them. People do not necessarily need to confront them, and thus feel vulnerable against them. In the West, however, in which people are demanded to dominate the world, these creatures with the supernatural powers represent a clear danger to the people, since they are uncontrollable. By being driven by their logic the supranatural powers of the Dragons are the ones the make them fearful.

Depending on the exact line of thought, the objective of harmony can be different. In Confucianism, harmony among human beings plays the predominant role, while in Taoism, harmony with nature plays an equally important role. In both cases, however, the notion of balance is considered to be essential, and thus the indicated route of choice is the "middle way." The resolution of disputes amongst Easterners takes greatly into consideration the context and attempts to satisfy both parties, frequently by compromise, while with the Greeks, logic dictates only a single truth that is derived by rationale. This truth is single and unique and does not change with the context. Correct is always correct while wrong is always wrong for Westerners. Thus, religious wars were quite uncommon in Eastern history but so common in the Western history.[16] Collective decisions in the West, traditionally, either directly represented in case of democracies, or at least they expressed in the cases of societies run by a single ruler or an oligarchy, the average of individual thoughts and the needs of the group's members. Of course, disagreement between the public feeling and the ruling authority frequently occurred and actually still does at various times and places; however, it was resulting in a distinct and identifiable oppression that eventually had to resolute as history continued. This "wisdom of crowds," which consists of free-willed individuals and constitutes the fundamental stone of democratic societies, is very nicely described and explained by Surowiecki (2004) in his homonymous book. In this book, though, special emphasis is given to the fact that the individuals that constitute the crowd or the group have to be independent, with minimal influence from the other members of the group. Thus, the corresponding decision is not collective, since the people are required to maintain their intellectual independence.

Consistent with these notions, Greeks advanced individualism and viewed the freedom to satisfy curiosity as means to reach the ultimate happiness, while in Chinese society, the cultivation of a collective conscience gains a major role, and the interaction with family, friends in a harmonious natural environment is central. The quest for happiness as the ultimate purpose of life is, in the West, achieved by the advancement of individualism, while in the East by cultivating collectivism. This advancement of individualism in Western thought is quite central and can be seen in various expressions of their civilization, even in music. While Western music is polyphonic, Chinese music is monophonic. Beauty in the West is achieved by the combined result of putting together successfully different independent musical instruments playing their individual music. Individuality is not lost within the musical group; instead, the music produced by each instrument contributes to the whole result.

An analogous principle applies to medicine. Western medicine, founded by Hippocrates of Kos, is quite analytic, trying to understand the cause of the disease and, subsequently, to intervene with that. More importantly it rationalized the pathogenesis of the disease dissociating it from contextual, vague and occasionally metaphysical forces. On the opposite side of that approach is Chinese medicine, which is holistic and attempts to restore the lost balance in the body, a balance that is maintained by the equilibrium between the yin and yang. Science did not advance greatly in ancient China because the satisfaction of curiosity was not a major issue, but technology did because it could help people and societies to overcome certain difficulties they faced and to achieve harmony. It doesn't matter for Easterners how an invention works, but rather that it works!

The advancement and domination of the Western world over the rest during recent human history is summarized by Niall Ferguson (2011) through the six "killer applications" of the West: competition, science, property, modern medicine, consumption, and work ethic. All of them attend quite well the legacy of the ancient Greek line of thought and are deemed responsible for the prosperity and predomination of the West.[17]

The cultural divergence of the West and the East may also explain why Chinese culture unified quite early, with societies soon organizing around a centralized government, while in the Greek civilization, the independent city-states represented the characteristic form of civil and political organization. Despite that organization of their societies was more or less similar, they believed in the same gods and spoke the same language, ancient Greeks identified themselves primarily by their city-state.

For the latter, of course, the mountainous geography in Greece did not favor agriculture much and also induced some isolation, but it is hard to attribute these great differences solely to this factor. After all, extended exchange of ideas, goods, and genes did occur in ancient Greece, which could have facilitated the generation of a single society organized under a centralized government. Apparently, the Greeks felt more comfortable in smaller structures that did not compromise much of their individual identity than in larger societal structures in which their individuality would have been diminished. This way their voice was heard more! Unification in Greece occurred soon after the Classical Period, during the times of Philip the Macedonian; unification continued and was profoundly extended by his son, Alexander the Great (the so-called first imperialist in global history), who reached as far as India. These periods were quite unstable, however, and did not last for more than a few generations, possibly because they were not reflecting the demands and the consensus of a society that has not succeeded in feeling unified. As soon as the strong central administration ceased to exist, dissemination of these political structures started to occur. The Roman Empire that followed, lasted longer, but again during its reign, it was very clear what the predominant culture was and serious attempts for cultural unification were not made (probably because it was well known that they would fail).

The notion of "field dependency" is central in Chinese culture and is in line with their agricultural tradition as well as their holistic view of life and their desire to search and rely on *relationships* rather than individual *objects and categories,* as dictated by the analytical approach of Westerners (Nisbett et al., 2003). Consistent with this view, an action or condition cannot be judged as right according to the Chinese without taking into consideration the specific context, as defined by the whole frame of relationships with other people, society, and physical environment. This notion is very nicely illustrated in the following old Chinese story, taken from Nisbett (2003):

> "An old farmer had a horse that ran away and his neighbors came to commiserate with him. 'Who knows what's bad or good?' he replied to them, refusing their sympathy, while indeed, a few days later the horse returned bringing a wild horse with it. Again, the neighbors came but this time to congratulate him. They received again the same reply from the old man: 'Who knows what's bad or good?' Indeed, his son, after a few days, tried to ride the wild horse but broke his leg. The neighbors again came to express their sympathy for the misfortune of the old man's son and received again the typical answer from him: 'Who knows what's bad or good?' Again, the village went to war some time later but the son was unable to participate because of his accident."

The story continues in the same motive, according to which the same event or condition cannot be judged as negative or positive without taking into consideration the context in which it appears. Something cannot be good or bad as such, but rather good or bad through the specific conditions in which it is examined. This is also in line with the farmer's "seed and soil hypothesis," according to which a specific seed is good only when planted in fertile soil that frequently is appropriate for that seed exclusively. There is a specific relationship that signifies the efficiency of the outcome that, in turn, does not rely only to the identity and the properties of the specific objects. Thus, the predominant for the Easterners' concept of the whole, the notion of collectivity, gains increased significance, and its value is determined by the quality of the relationships, along with that of the individual elements. The "obedience to the elderly" of Confucius also underscores this, since it places the decision of what one should do, from the level of judgement and evaluation, to the level of following the actions that will re-enforce continuity.

The fact that Westerners are more attracted by the specific individual objects and their particular properties offers a preference in individualism at which the value of the relationships appears second to that of the elements involved. After all, the atomic theory of matter, according to which the whole world is composed of invisible and tiny particles, is credited to the ancient Greek philosopher Democritus who intuitively developed his theories, not because he interpreted experimental data, but because he felt that this reflects more accurately how the world works!

The perception of collective consciousness, which is highly esteemed by Easterners and is surpassing that of individual consciousness, has dictated the whole organization of their lives, even until today. Elements of this approach can be identified even to domains of life that are, at their essence, considered highly individualistic, such as issues related to economic success. An example is offered by the recent success of the Chinese economy that has occurred after Western practices have been incorporated in industrial production. This success, and the related to that positive projection (the Chinese giant in industrial production as usually called), is likely that it is also due to this exactly collectivistic way of thinking of the workers. And this is not a conscious decision, but an unconsciously followed norm that is attained just because this is the way things are. According to this notion, the company's productivity and well-being is dominant against that of that of the individual workers'. Thus, wages can be kept at the minimum for more prolonged time periods without posing an imminent danger to the stability of their society. Characteristically, according to Ferguson (2011), the average Korean works 1,000 hours more a year than the average German! Of course, in that case, the question is for how long this can continue until Korea, China or other Eastern Asia countries abolish their competitive advantage. Economists will probably argue that this cannot continue for long, and that such collectivistic attitudes will retreat with a buildup of a liberal capitalistic society. Eventually, it will have to give way to several independent prospective businessmen and investors who, after a certain point and as soon as the wages of the workers and overall quality of life rises above a certain threshold, will start demanding a larger percentage of the profits. Instrumental to this transformation is the ongoing globalization at which Western-like behavioral patterns and norms are readily accessible to Easterners (and vice versa, of course). As we'll discuss in subsequent chapters (12 and 13), analogous are the predictions if we focus on behavioral trends.

The recent COVID-19 pandemic and how China and the United States dealt with it is another example that illustrates the difference between the Far East and the West. When it originally appeared in China, at the end of 2019, it spread in Wuhan Province and soon spread to the rest of the world. Soon, however, China contained it, and by March 2020, new cases were steadily below 100 per day, while by the end of 2020, they were less than 10. However, in the United States, COVID-19 continued to surge with more than 200,000 new cases per day in December 2020. Several reasons account to this difference. All of which can be reduced to the measures taken by governments and populations' adherence to them. Therefore, this difference exemplifies the distinct attitudes of the two societies; the individualistic in the U.S.A. and the collectivistic in China. The U.S. government was reluctant in taking harsh measures to contain the pandemic because they will oppose the individuals' freedom and they would not be received well. It has to do with the economy, the businesses, and the idea that it is up to people to protect themselves from the virus. In China, though, people were more obedient in practicing more extreme measures, at the expense of their individual freedom and the pandemic was contained. Of course, in China, the laws were enforced more strictly, but again, it is ultimately up to the people to be perceptive for such severe laws that may restrict their individual liberty.

Other facets of this behavior, beyond the purely (if there is such a thing!) economic level, might also be related to more political issues. Drastic changes in the political systems or the ruling parties, were limited in Eastern societies, as compared to those of Western societies, a trend that was also reflected by the many different nations and states within the European continent that follow the political tradition of the ancient Greek city-states. Even more striking is the frequency by which borders changed during the later European history, as well as to the strong nationalistic feelings that persist even today within the European nations. Virtually all citizens of European countries feel quite distinct from the citizens of the countries they share borders with, and not infrequently, even from other citizens of their own country! Being European is quite secondary to being French, British or Spanish. In the U.S.A., along with the American identity, the "state" identity also plays a major role, which may reflect the strength of individuality. During the Roman era, that Europe and its surrounding territories were unified under a single administration does not contradict the Balkanization tendency of Westerners, since who the ruler was and who the one that has been occupied by force, remained clear throughout history. Despite their multinational populations and their geographically extended borders, in all examples of European Kingdoms, which one was the subordinate, and which represented the dominant population was always quite clear with the concept of a federation of equal Nations being rather vague to non-existent.

Finally, today we all witness the difficulties of the attempts toward European unification, and only a few years after the introduction of the common currency, it appears how unstable and vulnerable this still is. About ten years after its original introduction, around the year 2,000, a chief concern of whether the Euro, the common European currency, had a future in history emerged. Several countries started doubting its financial value, and all European countries, 20 years later, have political parties that structure their political platforms and are actively advocating for their exit from the Eurozone.

As opposed to the Europeans, the Chinese, throughout their history, remained a largely unified society that was mostly ruled by a single administration, without strong – or at least consistent - attempts for territorial expansion. It is conceivable that it wouldn't be like that if the perception of the collective consciousness did not surpass that of the individual.

Several similarities of Eastern-collectivism and Western-individualism can also be identified with political systems and ideological debates that highlighted, especially the 19th and the 20th centuries. Today such disputes are not as vibrant as they were in the recent past, yet communist rule persisted in China but not in Russia and Eastern Europe. However, in Europe and the United States, where they associated with individualism, capitalism thrived.

Such "collectivistic" or "individualistic" decisions are not related to a notion of right and wrong, and by no means does a rule of majority apply that may imply that the collective is wiser than the individualistic. It's just differences between perceptions and life approaches. After all, Surowiecki (2004), in his *Wisdom of Crowds*, shows that the crowds are wise only if they have achieved the highest degree of independence, while interdependence reduces this wisdom!

61

The concept of balance or complementarity between sets of characteristics reflects actually the establishment of different "survival" strategies and is not new in biology. For example, some species produce several offspring to increase their chance of survival for some of them, but they do not take care for each individual one of them a lot. In these species new offspring usually become independent quite early in their lifetime. Others, however, produce only a few each time but invest a lot in their survival. In these species, usually, the offspring stay with their parents for more time and are taken care by them, protected and fed. People are like this.

Beyond biology, we apply such strategies intuitively in our daily lives as well. Some prefer to take care of our savings alone, without professional help, while others reduce slightly their potential returns by sharing them with professionals, but with the expectation that the portion that will return to us will not be lower from that if we did it alone. Alternatively, we prepare our tax returns alone, at the expense of our own time while others do it with the professional help. This incurs expenses but allows for more time that can be invested in our own professional activities. Furthermore, it assures that mistakes will not be made, and penalties will not be paid. It is the cost of relying on someone else, an expert, that assumes a part of our responsibility.

Such strategies are also applicable to whole countries, with some investing exclusively in technology and others both in technology and food production. The former is not sufficient in terms of food supply but the revenue generated by the technology products will allow them to purchase food from elsewhere. And this can go on to every aspect of our life. It is all a matter of the strategy we chose to follow. It is not right or wrong, but it is just a matter of how well and efficiently it is being practiced in achieving its goals. This applies to the fundamental differences between different cultures, what they want to achieve and how effectively they are in doing so. And the question we will try to address, of whether different people are "hardwired" for one or the other.

So, collectively, during human history, the people who occupied and inhabited the corresponding areas developed two major lines of thought, the Western and the Eastern, with fundamental differences in the way life is viewed, explained, and eventually lived.

CHAPTER 6

Population Trends versus Individual Traits

The assumption that populations that have genetic pools with different allelic frequencies at certain genes—and that this can affect behavior—is based on an oversimplified extrapolation that, in turn, bears certain limitations. It is wise to keep these limitations constantly in our mind during this discussion.

The most apparent such extrapolation is due to an arbitrary, however unavoidable, jump from the individuals towards the collective population's traits. In other words, that a certain trait for a given society becomes more intense if more individuals with this specific trait are present in this society. Simulations on the behavior of individuals as members of a group as well as the effects of genes that act at the level of the group (instead of the individual) may prove the limits of this assumption. However, in order to appreciate the magnitude of these limitations, let's attempt to view this by a more simplistic and qualitative manner. To that end, let's assume that there is a polymorphic gene that can dictate an easily manifested behavior, such as an aggressive behavior. We don't refer to aggressive behavior in that case to its formal psychological, more strict definition, which may also be associated with violence, but rather to its widest meaning, which reflects the tendency to impose a certain way of life and reactions on others and to resolve possible disputes by force and confrontation. This force can be both physical as well as psychological and intellectual. First of all, this hypothesis is oversimplistic per se, since there is not such a gene that affects behavior and controls a single trait only. As we will see in detail in the following chapters, in all cases, a single genetic polymorphism affects more than one behavioral trend, and psychologists find this quite reasonable since they view all such trends as interconnected.

In any case, let's now assume that this gene is represented in two forms, with allele A1 associated with more elevated aggression than allele A2. If A1 is more frequently present in group A, then this population contains more people hardwired for aggressive behavior than the other, the B population, at which A2 is more common.

Our simplistic assumption continues; population A has more aggressive individuals and will also be a more aggressive population as such, as compared to the population designated as B. Taking this hypothesis one step further, we may contact a hypothetical historical experiment in which the two populations, A and B, participate. All parameters in this experiment, including environment, population size, as well as the rest of the genes—excluding genes A—of the allelic frequencies are the same. In other words, the two populations are identical except the number of the aggressive individuals they include. The experiment starts when we take the populations A and B and we put them together, offering them distinct territories in close proximity to each other. We can even limit both their food supplies in order to promote and enhance confrontation. It is rather likely that we will soon witness a fight in which the A people will try to dominate—and possibly soon will—the B people. So, at the end of the experiment, and since all other parameters besides the A frequencies are similar, we may expect to see a single population with two distinct subpopulations, the A and B, with the former dominating the latter. Simply thinking, this is quite true and well anticipated, and reflects that in simple confrontations the more likely winner is the one exhibiting the more aggressive behavior.

It is more complicated, though, when we talk about populations instead of individuals, and this is not only because the system is more complex and unpredictable, but for a number of other good reasons as well. For instance, an alternative scenario would be that following their interaction, the aggressive A people should also manifest aggressiveness not only against the B people, but also against themselves. It is conceivable that this will have to do with the share of power (or resources) in the new group. After all, they are aggressive people! In that latter case, it is not unlikely to expect that the B people will be the ones that will dominate the mixed population, since the A people will be much too preoccupied to satisfy their confrontational instincts, while the others may be more able to dictate the terms of co-existence, thus appearing as the dominant population. This possibility becomes even more apparent if, at time 0, which signifies the initiation point of the "experiment" instead of putting the two groups instantly together, we leave them alone for a certain period of historical time. In that case, it is conceivable that the civilization of Bs will be more advanced, since we may expect that massive aggressiveness usually compromises collective advancement. At the end of this experiment, the independent observer, in that latter case, will record that despite that A people are more aggressive, it is the B people who have dominated the mixed population. So, at the level of populations, things may not come out the way we have anticipated as to the individuals.

Someone may argue at this point that if the successful outcome in such an experiment is measured as the degree of domination of the one population against the other, then some but not too much aggressiveness might be beneficial. Indeed, if aggressiveness is limited below the hypothetical threshold at which it becomes deteriorating for a given society and does not pose an obstacle to its advancement, this may be synergistic to domination. Thus, some but not too much aggressiveness may be good and a perquisite for success. Countless everyday examples, ranging from the behavior of schoolchildren among their peers, to the attitudes of various professionals, support this notion. So probably, this is a reason that robust cultures and civilizations were built and developed by populations at which the various genes we'll discuss subsequently differ in the ratios of their frequencies, but in none of the cases can we see absolute absence or presence of certain alleles. A plausible explanation for this is that the presence of the under-represented alleles is obligatory in some sense for the proper and calibrated functionality of a certain society.

We may envisage countless alternative probabilities and scenarios that result in the deviation from the result we originally anticipate. Nevertheless, and for our specific purposes, we may expect that a society with more (but not exclusively composed by) aggressive individuals will constitute a more aggressive and prone to domination society, or a group of people with many risk-takers and novelty-seekers will more frequently undertake exploratory activities and will be prone to change, and so on. What is interesting, though, in that case is that several of these tendencies are controlled by the same polymorphic markers. And even more surprising, probably, is that even those that aren't controlled by the same genes seem to go well together in the same populations!

Other phenomena also complicate the linear transition from the individual to the group. For example, the way social networks are organized and connected clearly shows that over the individuals, networks have a life

65

on their own (Christakis and Fowler, 2011). By being under the influence of their friends and acquaintances, people may start developing traits that are not only due to their biology and genetics. Conceptually, this likely operates as a multiplier for traits already present in the population, which in turn, points to the power of the quantitative factor, in allelic frequencies, as a multiplier for various phenomena. Let's take aggression, again, and assume that it is under the influence of both genes and the social network or friends. The non-aggressive people will eventually become aggressive by interacting with their aggressive friends. And the more these friends are, the faster aggressiveness in the society as a collective trend, will spread and be established. It is more interesting to examine this from the opposite side; with regards to the conversion of aggressive to non-aggressive people. In that case we may start appreciating the impact a collective culture has on individuals against traits that may be dominant in a specific context but may become recessive in another.

The strong dynamics of social networks and that they are also under some biological regulation was also demonstrated by the fact that genetics seem to play a role in our decisions of who our friends are (Fowler et al., 2011; Christakis and Fowler, 2014). By screening a large number of people for social relationships, the studies identified a significant tendency in people to pick friends from a specific genotype, similar to their own.

PART II

Personality Traits at the Population Level

CHAPTER 7

Exploratory Activity and Novelty-Seeking: The Case of Dopamine Receptor D4

Two of the earliest studies associating certain genes with a specific behavior were conducted almost 25 years ago by Hamer and co-workers (Benjamin et al., 1996) and Belmaker and co-workers (Ebstein et al., 1996) in an attempt to identify a genetic predisposition toward novelty-seeking. This trait is thought to have a certain genetic component and is associated with behavior that is impulsive, exploratory, fickle, excitable, quick-tempered, and extravagant. On the other hand, individuals who are not characterized as high-novelty-seekers are those who score lower than average in the corresponding tests and tend to be reflective, rigid, loyal, stoic, slow-tempered, and frugal. As happens with this type of study, the researchers examined the frequency of the alleles under investigation in groups of people who had been classified according to the characteristic under study, being the novelty-seeking behavior in this case.

The aforementioned genetic analyses performed by these research groups found an association of the novelty-seeking behavior with certain alleles of the gene that encodes for the dopamine receptor D4. Specifically, the novelty-seeking behavior was associated with the 7-repeat allele (see below). Not surprisingly, the results of these studies, like the vast majority of genetic association studies, should not be considered as definitive, since other investigators failed to reproduce the results (Paterson 1999); however, even though the statistics are not as strong as they could be, a certain element of truth or a trend must be there.

How does dopamine work? Dopamine neurotransmission is essential in regulating behavior, controlling, among other factors, movement, emotional responses, and one's capacity to feel pleasure and pain. Early antipsychotic drugs acted by inhibiting the activity of dopamine, underlining the significance of this system in regulating behavior. Dopamine is a neurotransmitter that, upon binding to its corresponding receptors, including the D4 receptor, inhibits the levels of a second intracellular messenger known as cAMP. D4 receptor in particular is polymorphic, exhibiting alleles that differ in the repetition number of a certain stretch of DNA consisting of 48bp nucleotides. Thus, depending on how many times this 48bp-variable number tandem repeat (VNTR) within the D4 receptor is repeated, different alleles exist in the population. The most common allele in the general population is considered the one that has four such repeats, followed in frequency by the one with seven repeats.

How is the number of those repeats linked to the different properties the D4 alleles may exhibit? At the functional level, a difference in the activity between these alleles has been found. Specifically, these alleles are thought to differ in their ability to reduce the levels of cAMP within the cell, with the one with the seven repeats being less active than the ones with the four or two repeats. Therefore, cells, and thus individuals, bearing the 7-repeat alleles fail to have the cAMP levels efficiently reduced as compared to other individuals with more active DRD4 alleles, such as the shorter four- and two-repeats alleles.

In terms of the evolutionary history of the DRD4 gene, the specific potential "value" of the 7-repeat allele is also pointed out by the observation that—contrary to the other DRD4 alleles that are considered as simple, one-step, molecular derivatives of a single allele—the 7-repeat allele differs from other alleles by at least six molecular events (Wang et al., 2004; Ding et al., 2002). This indicates the operation of positive selection pressure for this allele that resulted in the increase of its frequency in human populations. Thus, individuals bearing this allele had an advantage (or in cases of group selection, the advantage occurred at the group these individuals belonged to).

7.1. DRD4 Alleles Have Different Frequencies in Different World Populations

Interestingly, different alleles of the DRD4 gene, along with a differential activity, also show quite different distribution among different populations: For the 7-repeat (novelty-seeking) allele, it was found that it was much more common in Europeans and populations of European origin, such as the Americans, as compared to Asian populations (Chang et al., 1996). Whether these differences represent the collective and cumulative result of selective pressure or they are due to founder effects related to the genetic composition of the early populations that inhabited the corresponding areas remains elusive and is actually impossible to prove or disprove with certainty. The fact, however, is that the allelic frequency of the novelty-seeking, 7-repeat allele for most of the Europeans studied, ranges between 0.06–0.21, while the corresponding value for the South East Asians was considerably lower, approaching 0 for most of the cases (Table 2). In this table, the sum of the frequencies of the 7-repeat and

Table 2. Allelic frequencies of the 7-repeat and 4-repeat alleles of DRD4 in selected populations (adapted from Chang et al., *Hum Gen 98*: 91, 1996)

Population	4-repeat allele frequency	7-repeat allele frequency
Africa		
Biaka	0.76	0.14
Mbuti (Zaire)	0.83	0.16
San Bushmen	0.91	0
Falasha	0.83	0.11
Europe and Middle East		
Mixed European	0.57	0.21
Roman Jews	0.63	0.19
Danes	0.67	0.14
Finns	0.69	0.06
South and East Asia		
Chinese Han	0.76	0
Japanese	0.79	0.01
Cambodians	0.52	0
Yakut	0.83	0.04
Malay	0.42	0.17
South America		
Colombians	0.23	0.62
Ticuna	0.20	0.78
Karitiana	0.39	0.50

4-repeat alleles is not equal to 1 because additional, rarer alleles have also been identified in these populations. A noteworthy exception in this distribution is the corresponding frequency found in Malays that had this allele at a frequency as high as 0.17, deviating from that of their Asian neighbors and approaching or even surpassing the range of frequencies recorded for the distal Europeans.

Omitting, however, the details, we will see that Asians have in general a lower incidence of this 7-repeat allele than Europeans.

Let's now recall the fundamental differences between the mentality and the perception of life between Easterners and Westerners, and then try to understand them in view of the characteristics that underline the personality of the novelty-seekers described earlier. In general, Westerners appear to be more independent as persons, having the concept of individuality as a fundamental element of their civilization. Their increased excitability and attraction to extravagance that are intrinsically linked to the novelty-seeking behavior, are also related to this individualism and the tendency to constantly seek for means to obtain satisfaction. The Westerners are certainly more keen explorers, as well. After all, they were the ones who actively and repeatedly attempted to meet the Easterners, and not the other way around.

Continuous exploration at all levels, including physical, geographical, and intellectual, is an undeniable feature of Western civilization. Particularly in today's Western and Westernized societies, such exploratory activities are not performed in order to satisfy needs that developed, but they emerged as an actual need that has to be satisfied.

On the other hand, Easterners appear to be more passive against life, an attitude that reflects a certain degree of stoicism and makes life within larger—and likely collectivistic—groups of people more convenient. Taking things the way they are probably reflects their belief that there is not much that one can or should do to change them. This is probably the reason that these people appear rigid against life and loyal, a fact that is also reflected historically in their relatively high political stability. Although Easterners are undeniably hardworking people, their more intrinsic motivation is probably not related to a notion of individual advancement, such as in Westerners—a notion that in turn has assisted capitalistic and individualistic societies to flourish (and, vice versa, flourished within these societies). It is instead related to a sense of loyalty against what should be done, what the regulations and the establishment dictate, and of what should be done according to what is thought to be the common and collective good, for the benefit of the society as a whole.

This attitude against life is also probably assisted by the frugality exhibited by the non-novelty-seekers, like the Chinese. A likely consequence of being frugal is that no matter what your individual productivity is, a larger—in general—portion of your product may go for a common cause than return directly to you, the producer, as a profit. Therefore, it is not the individual but rather a larger sum of people, or a society, that receives such benefit, while the individuals only indirectly benefit, just by being parts and components of this society. This in turn strengthens the collectivistic behavior.

It could be argued that today's well-organized societies of the West try to cultivate (and indeed have managed to) develop an increased conscience of the common good, while they have also nurtured the concept of discipline that is a perquisite for them to operate. It is astonishing how intensively the service to the society is nourished in the individualistic United States. As early as in primary school, kids are prompted to perform service activities to count favorably in their future.

These all are certainly true; however, it is possible that such tendencies, instead of reflecting a holistic view of the world, are an unavoidable specific requirement for such societies to function and to allow individualism to further develop. As opposed to this notion, the concept of rigidity and loyalty against social structures and hierarchies, that is intrinsic to Easterners, is also in good line with the behavioral characteristics and attitudes of those that are not novelty-seekers.

Easterners, with their increased context-dependency, view happiness, not as a state that is remote from the "whole," but deeply associated with it. You don't need to go away to seek happiness; it is here with you as long as you can discover it and reveal it. Such an attitude is inherent in societies at which collectivism tends to surpass individualism.

Today, we can all see that in China, in one of the fastest-growing economies of the world, the average standards of living are much lower than those in the West, and that those standards may be disproportional to contemporary Chinese productivity and economic development. Of course, this in part is due to very recent political and historical reasons related to the fact that only recently has China adopted some more or less typical capitalistic practices while coming from a totalitarian regime. However, even in neighboring Japan, at which the capitalistic mentality is more mature and comprehended, the collective benefit frequently prevails. Probably in that case the larger structure we refer to is not the whole population of these people. It can also be the corporation (the employer), but again the individual good gives its place to the good of something larger and greater than that of a single person. R. Nisbett, in his *Geography of Thought,* offers several nice examples describing in detail these differences.

The fact that Eastern science is not, in general, considered as innovative as Western science—although it is certainly more technologically oriented—can be explained by the fact that while Westerners, by being better novelty-seekers, tend to discover and explore. These characteristics are tightly linked to the advancement of science. Advancing science in the West does not represent means to reach a specific goal but is rather due to the satisfaction of individuals' curiosity on understanding how the world functions. The fact that specific results of scientific research are frequently useful and applicable is not an integral subordinate of the investigation process but rather a desired outcome. The actual motive is to understand how the world works. That's how science and scientific inquisition has been structured historically in the Western Societies.

Contrary to that perception, Easterners are more prone to technological advancements that ultimately attempt to satisfy needs of society, and are not dictated by plain curiosity. "If an invention works, then it's not necessary to know how this is achieved" can be the motto of this technological advancement. Alternatively, who cares how the world works if this doesn't make our lives better? The recent investment of China in basic science doesn't contradict this approach.[18] It just shows that we may have reached a point that basic and applied science are practically indistinguishable, as well as to the fact that this may be driven by other motives related to the nation's prestige in the global scientific community. Nevertheless, in any case it is mandated following recognition of specific needs.

7.2. DRD4 and Human Migrations

The comparison between Easterners and Westerners, as regards the novelty-seeking behavior, appears to be, in general, in good line with the reported data on the frequencies of the corresponding alleles, with the Eastern populations exhibiting lower frequencies of novelty-seeking alleles as compared to Western people. If, however, we look at Table 2 more carefully, we'll notice some interesting twists. We'll see that the populations from South America have very high frequencies for the 7-repeat allele, surpassing that of all other populations examined. A high novelty-seeking behavior, by the cultural standards described earlier, is not sufficient to explain this for the South Americans, since they haven't exhibited evidence for high exploratory activity, at least as viewed according to typical Western eyes. Of course, the existence of certain historical reasons that may account for this unexpected observation explain very satisfactory why these cultures and civilizations did not show it, along with the degree of scientific advancement we may have anticipated to accompany the very high frequency of novelty-seeking alleles in these peoples.[19]

If, however, we employ a different view and try to trace the roots of this finding in the ancient history of these populations, and particularly their migratory history, we may find some additional explanations, as proposed by Chen et al., (1999) and Matthews and Butler (2011). These investigators discovered that there is an interesting correlation between the presence of the novelty-seeking DRD4 alleles and the physical distance of out-of-Africa migration in the corresponding populations. Roughly, they noticed that when longer distances had to be covered during these migrations, then they were associated with higher frequencies of the novelty-seeking alleles in the corresponding people. Consistent with that view, a plausible explanation becomes apparent. Between 20,000 BC and 10,000 BC, modern humans had colonized the entire world. In order to effectively perform these rapid migration attempts, among their other characteristics, they also needed to adapt into the new environment and also to exhibit low reactivity to novelty stressors. They had to be good explorers and they also had to be able to cope with great differences and variability when, during their trip, they had to change their habitat. Otherwise, they wouldn't be able to travel from their original sites in Africa, as far as in South America, via North America, through all these highly variable environments. Of course, it was not exactly the same individuals who left Africa that eventually settled in South America. It is conceivable that this trip lasted for a few generations and, in between certain subpopulations, quite likely settled in various locations. Probably, the novelty-seeking alleles were enriched by this process. Still, such exploratory activity had to be intrinsic to the people. Whether in that case the novelty-seeking alleles of DRD4 actually induced migration or they rendered the individuals that bear them feel more comfortable in the new habitats remains is debatable, with the second possibility appearing as the more likely. Nevertheless, the fact is that these particular people who originally colonized South America needed, or were assisted by, at the least, a specific genetic signature.

Thus, it is quite likely that the specific behavior induced by this genetic characteristic, that of novelty-seeking, was not as clearly demonstrated in South Americans as in Westerners, as we have seen by their prior comparison with Easterners. This can be due to the strong effects of special socio-environmental and historical reasons, or alternatively to the consequences of other genetic alleles that might have masked these effects.

7.3. Nomads and Settlers

That the 7-repeat DRD4 allele is present in Africa at variable frequencies is not surprising, considering that Africa, is the source of all human variation. Thus, the populations that have departed from Africa carried the polymorphisms that were already present in the populations from which they originated. However, some interesting observations that offer additional value to the potential role of this gene in affecting human behavior and, ultimately, history can also be found within certain populations in Africa that differ in the 7-repeat allele frequency.

Ariaal people are essentially nomads who wander around northern Kenya. Between 1960 and 1970, some of them, under the influence of Christian missionaries in the area, settled in certain locations, establishing societies that practiced agriculture. This represented a very strong and acute change in their way of life that, until that point and for several generations before that, was nomadic. This was also a very recent event that provided a very "informative" tool for anthropological and other studies. Eisenberg and co-workers (2008) took advantage of this naturally occurring "experiment" and examined two groups of Ariaal men that lived in Kenya. The first group consisted of the people who were still nomadic, while the second group included Ariaal people who, for approximately the last 35 years, lived in villages. Within that time frame, no considerable changes in the genetic signatures of the corresponding populations were anticipated, since the time interval did not surpass that for one or two generations. Then they studied the frequencies of DRD4 polymorphisms in association with various nutritional indices that are considered indicative for better adaptation to the nomadic or the settled life.

The investigators found that the presence of the 7-repeat allele is associated with better nutritional status in the nomads but worse in the settled individuals. Thus, the nomads with this allele appear to possess an advantage as compared to those that do not have it, or the settlers that also have it (Figure 5). The results of this interesting ongoing evolutionary experiment, in terms of selection pressure and consequences in these populations, will be recorded in the future.

These interesting implications of DRD4 in the lifestyles and behavioral patterns become even more complex in view of certain observations linking the same DRD4, novelty-seeking and "nomadic" seven-repeat allele, with the childhood onset psychiatric condition designated as attention deficit hyperactivity disorder (ADHD), and for which the major symptoms are inattention, hyperactivity, and impulsivity (LaHoste et al., 1996). It is actually not hard to identify specific similarities between ADHD and

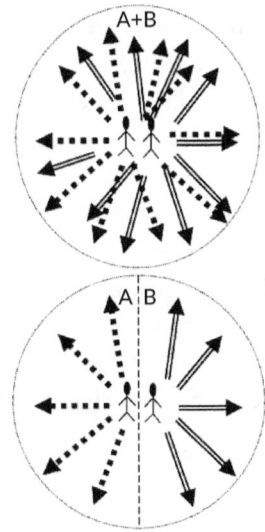

Figure 5. Nomads *vs.* Settlers in food-gathering. Nomads (top), have to collect food from any possible location independently within a defined territory (A+B). Settlers (bottom) have divided the territory into sections A and B and co-operate by sharing the food they collect. Nomads are more independent and self-sufficient while frequently they have to cope with various novel stimuli. Settlers are more interdependent and they can afford developing a routine in their daily lives. Dashed and double arrows indicate food-gathering moves of each individual.

the novelty-seeking behavior. Apparently, the latter, under the influence of certain environmental factors or even that of other genes, may cross the limits that define "normal" and turn into pathologic (notwithstanding that the line between normal and pathologic is as always quite vague). So, the same gene that is related to a pathological condition such as ADHD also offers an advantage to nomadic life. While for a pupil in a western-style school being hyperactive and impulsive is a disadvantage, for a young Ariaal nomad it may be advantageous, as he will be of assistance to protect more efficiently and feed himself better.

Even though nomadic life stopped being as common as in the past, we may view possible remnants of such behavior in contemporary societies, particularly if we extend the concept of "moving from one physical location to another" to other attributes of daily and social life, such as concepts, beliefs, and ideas. For example, the attraction of Easterners to relatively large (and interdependent) communities, with their great attachment to tradition and things inherited by their ancestors, may reflect a settler's approach to life. Furthermore, the increased context-dependency, or in other words, the fact that they always view and understand themselves as parts of a wider whole, may also be related to that notion. This wider whole may not necessarily be restricted and defined by certain physical boundaries but can also be extended to ideas, social structures, and relations, and generally attitudes towards life. The settler's approach implies that someone lives and dies within certain and well-defined borders, not only physical but also intellectual. These borders define the frame in which someone's life develops, and in traditional collectivistic societies it is quite likely that they have changed little between generations.

On the other hand, Westerners, perhaps, are not typical nomads, and actually many of them never were; however, they are still attracted by attitudes possessing the common denomination of "wandering around," which is an intrinsic feature of nomadic life. For example, while they have—and actually, concomitantly with the development of Western civilization they always had—developed a special relation to their home city or country, still, intellectually they are attracted by various changes, they easily doubt traditional values, and generally they are skeptics and critical against things given as such. The latter is actually a mandatory condition for the development of a rational scientific thought and predominated all aspects of life in Western history.

Furthermore, the whole concept of individualism mentioned earlier may also represent an expression of certain facets of the nomadic life in their widest sense. Nomads do not have an attachment against a specific physical location. By being predominantly hunters and gatherers, and by possessing herds they take with them in order to satisfy their needs for food, they exhibit reduced interdependency with each other. In parallel with that notion, people in an individualistic society attribute increased importance to self-reliance, thus trying to reduce their dependency upon others. Money usually provides the media to do that today. And since in contemporary complex societies interdependence is very high, the preservation of individuality is maintained and satisfied by the rationalization of this interdependency and by establishing detailed laws and regulations that protect from the deterioration of society.

A remnant of this nomadic attitude can also be the little attention typically given by Westerners to the effects of their civilization on the environment. Similarly to the typical nomads, who as hunters and gatherers don't care much about the effects of their immediate actions on the environment, as long as they satisfy their needs (nature will find its way), the contemporary Westerners started to worry about the environment only recently, when they realized the catastrophic consequences of their activities.

7.4. A Gene for the Liberals

The saga of the direct implications of DRD4 polymorphisms and the organization of social and actually political life goes even further, beyond the establishment of nomadic or settler's life. In 2010, an intriguing study was published by Settle and co-workers (2010) in the *Journal of Politics*. In this study, led by senior authors N. Christakis and J. H. Fowler, authors of the classic bestseller *Connected* (2011), the investigators claimed that the presence of DRD4 7-repeat alleles is associated with political ideology, specifically liberalism. By examining about 2,000 individuals, the authors discovered that by possessing the novelty-seeking allele, the number of friendships a person has in adolescence is associated with liberal political ideology. In other words, the novelty-seekers were found to be more "receptive" to the exogenous variable signals that may be expressed in the acquisition of liberal ideology. In this study, the subjects were asked the simple question: "In terms of politics, do you consider yourself conservative, liberal, or middle-of-the-road?" and had to select among the following equally simple answers: "very conservative," "conservative," "middle-of-the-road," "liberal," or "very liberal." The authors explained their findings by hypothesizing that "[I]ndividuals with a genetic predisposition toward seeking out new experiences tend to be more liberal, but only if they are embedded in a social context that provides them with multiple points of view." Thus, having many friends and being exposed to various different ideas may affect your political ideology, provided that the genetic background is sufficiently receptive, as defined by the presence of DRD4 7-repeat alleles. In other words, a specific genetic fingerprint determines how prone one is to the influence of many different ideas, a fact that may be sufficient to make him a liberal. These investigators finally concluded to the ground-breaking and yet completely rational statement that, "In light of these and other findings, political scientists can no longer afford to view ideology as a strictly social construct."

Naturally, we can only imagine what the result would be if an analogous study had been performed on individuals from a society of a different political system, especially such as those that existed during the days of the Cold War until a few decades ago, and for which conservatism and liberalism had a completely different meaning. In any case, though, a specific genetic polymorphism has been linked to political ideologies.

If we turn again to Table 2, we'll see that a nice correlation exists between people having the novelty-seeking, 7-repeat allele and political history that is rich in revolutions, drastic or not so drastic changes in the organization of their society and its structures, and (in general) societies at which political endeavors appear to play a major role in daily life. We will readily realize that the people who follow the legacy of the ancient Greeks, namely Westerners, and for which the novelty-seeking allele is found at higher frequency than in Easterners, are those who continuously seek for more efficient means of government, challenge their existing governments and it is also quite common for them that, during the last years of their history, a given generation has lived under quite different political conditions than the previous generation.

As opposed to them, Easterners are the ones that, by viewing their lives as an integral part of greater whole, are reluctant to perform drastic changes; they are, in general followers, of the notion of a general stability in terms of a balance—facts that can be interpreted as an inherent attraction to what we interpret today as conservatism.

As regards South Americans, a rather intriguing possibility that may contribute to the political instability that characterizes these nations is that due to the extensive interbreeding between the local populations and Westerner conquerors, novelty-seeking DRD4 alleles passed into their genetic pool enriching them. Therefore, contemporary South Americans, by being the product of interbreeding between the descendants of ancient South American people (that have the novelty-seeking alleles at quite high frequency) and Westerners, have probably an increased incidence of 7-repeat alleles which, and thus politically speaking, they have developed a tendency for liberalism that may also be reflected by their politically instability. The latter, of course, is heavily influenced by various additional socio-economical and historical reasons, but is possible that is also enhanced by their genetic makeup!

7.5. DRD4 and Financial Risk-Taking

The novelty-seeking behavior that is likely related to DRD4 polymorphisms and is characteristic for Westerners transcends various other aspects of daily life and is expressed in reactions and attitudes that are not imminently related to exploratory activities. It appears that financial behavior also represents another facet of this novelty-seeking behavior.

In January 21, 2010, U.S.A. President Obama, in a discussion of the possible causes of the economic crisis, stated the following: "This economic crisis began as a financial crisis, when banks and financial institutions took huge, reckless risks in pursuit of quick profits and massive bonuses."[20]

With this statement, the president pointed to certain risk-taking decisions as a causative factor for the crisis, indicating that certain behavior patterns may have dramatic consequences in global financial stability. The question is how inherent such decisions are and whether they can really be avoided. Economists, and probably psychologists as well, might argue that such risk-taking decisions are integral to the system, and the same applies to the subsequent crisis that eventually followed. There might also be, however, a genetic component that makes these decisions more likely, explaining that the "integral" component we refer to is not limited to the financial and economical patterns related to the system *per se,* but is also extended to the genetic features of the societies and the corresponding people that undertake these decisions.

Dreber and co-workers (2009) studied the potential role of DRD4 polymorphisms in financial risk-taking by using an investment game with real monetary payoffs. In this game, 98 males between 18 and 23 years old were given a balance of $250 and asked to chose a fraction of this amount for an investment. In case of failure, the invested amount was lost while in case of success, the amount invested was multiplied by 2.5 and returned to the participant. It was a coin flip that determined the outcome of the investment, thus having a probability of failure or success of 0.5. Analysis of the results showed that individuals with the 7-repeat novelty-seeking alleles for DRD4, on average, invested more money than the ones without these alleles, thus pointing to DRD4 polymorphisms as genetic determinants of financial risk-taking. Similar results were obtained by Kuhnen and Chiao (2009) in an analogous experiment that also identified another gene, related to the regulation of serotonin activity, as an additional regulator of financial risk-taking.

What happens if we compare the risk-taking decisions between Easterners and Westerners, or the ones with the low and high frequency, respectively, of the DRD4 risk-taking alleles? Apparently, strong cultural differences do not allow a formal and legitimate comparison. Even if we analyze the results of investment games-related experiments by racial criteria involving subjects of the same population, i.e., Americans of Eastern or European origin, again certain cultural differences will be unavoidable and would affect the outcome of our analyses. Psychologists, however, argue that there are certain differences that they also ultimately attribute to cultural differences and discrepancies in the philosophical perception of life. For example, in assessing the uncertainty in answers against questions of general knowledge (internally generated uncertainty) or likelihood for the occurrence of future events (externally generated uncertainty), investigators found that Asians repeatedly showed greater overconfidence than Westerners (Weber & Hsee, 2000). Consistent with this finding, comparisons of Chinese and American students in making financial choices between options with high or low risk, showed that the latter were, in general, more risk averse than the former (Hsee & Weber, 1999). Similarly, it was found that the perception of risk was lower in Chinese who were willing to pay, in financial experimental settings, more than the American subjects (Weber & Hsee, 1998).

82

These findings appear to contradict the higher-risk-taking behavior of Westerners who have the 7-repeat DRD4 allele at elevated frequency as compared to Easterners. Interestingly, this risk-prone attitude of the Chinese is limited to financial risks. In different experimental settings that assess other types of risk-taking, such as academic risk decisions (expressed as the choice between writing a paper on a conservative or provocative topic) or medical risk decisions (expressed as the choice to take a pain reliever with a moderate but sure or strong but variable effectiveness), the Chinese appeared more risk averse than Americans. The latter observations appear to be in line with the genetically predicted risk-averse tendency of Easterners and dissociate financial risk-taking from other types of "risky" behaviors.

The discrepancy in the response between the different types of risk-taking between Easterners and Westerners may reflect a different perception against monetary and, by extrapolation, against certain materialistic issues. It is quite possible to indicate the degree of self-reliance these people have towards different matters. In other words, in case of the "wrong" choice, the ease by which one takes the specific risk may imply how much help an individual is expecting to receive from his or her family, environment, or society. So, Easterners actually can afford to take such "risks" because their environment will support them in case of a mistake and will correct the outcome of the "bad" lack, that after all is part of their yin-yang world. In addition to that, it may also reflect the reduced value of money as the "buying power" in the context of achieving happiness in life. Therefore, those individuals are also willing to take higher risks because they feel that they have less to lose. Westerners are alone in life, and they have to deal with the outcome of their doings because their dependency on others is smaller. Thus, they need to be more cautious and careful.

Besides the ability of the Eastern cultures to reduce the harmful effects of risky choices, the perception of risk as such also differs between these cultures, which eventually affects the ease of taking the corresponding decisions. Thus, the difference in these risk-taking decisions and the confidence against the corresponding choices are likely related to the fact that Easterners have been trained within a rigid cultural system and are used to following traditions instead of criticizing them. Therefore, they fail to recruit evidence that disconfirm their original hypothesis, a fact that eventually results in giving them increased confidence in their decisions, thus their willingness to take risks is higher. On the other hand, Westerners from very early on in their lives are being trained to criticize and to confront. Collection of arguments is not limited to those that confirm but also to those that disconfirm their arguments, thus reduced confidence is more consistent with their perception of life. Therefore, their perception of risk is increased. In turn, while Easterners appear as risk-takers, they do not consciously feel that they actually take a risk. For the same choice, they feel confident and, thus, they can go one step ahead of Westerners. So, while the experimentalist is viewing them as risk takers, they do not feel as such. A choice that is considered quite risky for Americans is not that much of a risk for Chinese. This notion is also reflected to the gambling behavior at which the Chinese exhibit significantly less probabilistic thinking and riskier gambling decisions than Westerners (Lau LY, 2005). In other words, the Chinese actually do gamble while Westerners try to rationalize, to evaluate chances and eventually try to win.

Cultural differences complicate things even more. For example, success in business for the Chinese depends on several factors that in order of reduced significance are fate, luck, feng shui, accumulation of good deeds, and—only lastly—by knowledge (Pitta et al., 1999). Again, we see the predominant role of fate and luck in determining outcomes. For the rationale-based Western societies, apparently success should be mostly dependent on knowledge. This example illustrates nicely how external factors are important for the Chinese in determining a specific outcome. Thus luck, and consequently risk-taking has a different meaning between these two cultures which in turn, also affects gambling decisions. Furthermore, gambling is a very old and quite acceptable social tradition in China, a fact that obviously also affects greatly the perception of inherent risk. Contrary to this, gambling in Western societies is morally and ethically debatable. Finally, Chinese are not as proficient in risk-taking, especially in issues not related to financial choices. Therefore, if we were able to "normalize" behaviors and responses against the perception of risk, and subject individuals to risks that are perceived as equivalent, we might have expected that Westerners will display more risk-prone attitudes than Easterners. In agreement with that is the observation that Chinese are less prepared to accept certain risks than Australians, according to a study that evaluated risk-taking (related to industrial facilities, natural hazards, etc.) in these populations (Rohrmann & Chen, 1999).

After all, risk-taking, especially in financial matters, is a rather Western approach. Development and progress, as we understand it as Westerners, require making decisions, which in turn possess a certain degree of uncertainty regarding their specific outcome. Easterners, on the other hand, are attracted by stability and harmony, conditions that diminish risk-taking. And even for action that can be seen as risky in the short term, in the long run they are just a component of the continuous balance.

7.6. DRD4 and Substance Abuse

Behavioral and personality traits indicate preferences. How these will develop and integrate within a person's life and how they will manifest as specific tendencies depends on various additional factors starting from early life experiences and concluding with society as a whole. As such, novelty-seeking, manifested as the inclination to search continuously for new experiences is not an exception and may have consequences that can be good and beneficial, or at least neutral at the level of both the individuals and that of society. Alternatively, it may predispose individuals to harmful behaviors. Certain evidence, although not conclusive as yet, points to a potential link between the novelty-seeking alleles of DRD4 and drug abuse (Lusher et al., 2001). Furthermore, the contribution of various other genetic loci also affects the predisposition to addiction, which can mask and even surpass the effect of DRD4. However, to some extent, such a link is quite understandable, since the subjects constantly attempt to try new experiences that ultimately may turn into drug addiction. Usually curiosity drives drug use in the beginning.

A fair comparison of the trends of drug abuse between Easterners and Westerners is almost impossible. Socio-economic and cultural factors greatly affect the outcome of such analysis, as well as the fact that setting the threshold of socially unacceptable addiction and accuracy in recording the results of such analyses might be quite different between these societies. From what we know, however, the prevalence of such phenomena were quite high in both Western and Eastern societies. The drug problem in the West is quite self-evident in the modern world, since statistics indicate that increasing numbers of (especially young) people are drug addicts. In the Far East, the Opium Wars in nineteenth-century China, when almost the entire army was addicted to opium, combined with the increasing trends of drug addiction in China today, also show that the problem is at least equivalent among societies. For these phenomena, apparently, the contribution of the socioeconomical environment plays a role that surpasses the role of genes.

One can argue that while in the West the relative ease at the availability of drugs, in combination with the demoralization of the societies and the abolishment of the role of strict values in the youth, are primary causes of substance addiction, while in the Far East, besides the side-effects of contemporary modernization, causes also included the function of drugs as an outlet against massive oppression.

7.7. Summarizing on DRD4 As a Prototype Westerner's Gene

Among the various genes associated with a predisposition against specific behavioral patterns, DRD4 is probably the one that exhibits the most striking differential distribution among Easterners and Westerners. This is not only due to the clear geographical pattern of the corresponding allelic frequencies, but also due to the variety of specific behavioral traits associated with it.

In summary, the 7-repeat allele of this gene is frequently found in Westerners but not in Asian people. This is in line with certain behavioral characteristics these populations (and cultures) have. Specifically, the 7-repeat allele has been associated with elevated novelty-seeking behavior and, associated with it, exploratory activity, increased human migrations, and a tendency towards (or better adapted to) nomadic life. It has also been associated even with liberal (as opposed to conservative) political ideology and perceptiveness to the impact of social networks and connectedness. All these features indeed tend to characterize quite accurately Westerners and the cultural norms they developed, posing the intriguing possibility that DRD4 can actually represent a single gene that can highly "predispose" for what we understand as the stereotypic Western-type behavior. Thus, we could imagine that an individual bearing the 7-repeat allele functions more efficiently in Western society while the one without this allele would probably be better suited to a society with Eastern-like structures. Alternatively, we could consider these as an indication that a society with more individuals bearing the 7-repeat allele is more likely to have followed historical lines and choices more typical of a Western society, while a population with a lower number (or practically deficient for this as it seems to be the actual case with Easterners) of individuals with the 7-repeat allele would more likely attend the collective historical outcome of Easterners.

Even though we assumed that all these genetic association studies were indeed true and absolutely accurate, and that all the proposed associations were reliable and predictive for the corresponding behaviors in other populations as well, there is a specific point that has to be kept in mind. As already discussed earlier, attempts to extrapolate from individuals' characteristics to those of a group of people and societies, posseses certain dangers and specific conceptual limitations. Let's take novelty-seeking and exploratory activity as an example. Does having too many novelty-seekers in a society necessarily mean that this society will be science and exploration oriented? I doubt it. What we currently perceive as a Western-type characteristic is the collective and balanced outcome of frequently opposing behaviors while domination of one against the other, instead of "pushing" against a specific direction, may have led to some disorganization and eventual instability. Furthermore, there is not a single behavioral trait associated with a single genetic locus alone, which means that the presence of a specific allele may modulate the likelihood of more than one behavioral pattern. In other words, too many novelty-seekers, most likely, would have the nomadic characteristics to prevail and therefore, to prevent collective decisions and conscience to develop. The latter is a mandatory condition to exist in a society in order to progress in science (and not only in that) and even to proceed towards exploratory actions. Thus, ratios matter and frequencies are not directly proportional to the trends observed.

This becomes even clearer if we take political ideologies as an example. We can assume that in certain societies the presence of too many liberal-minded individuals, in terms of arguing against and criticizing given and widely accepted values and norms, bears certain dangers when evaluated at the level of the whole group of people. It is conceivable that they occasionally will be dysfunctional and unable to operate collectively as a single society. Thus, these effects, to make a society efficient they should be accompanied and "balanced" by another trait, conferring specific obedience against laws. This way, controversies will be integrated into the society without compromising its integrity.

An actual observation that it is quite common in the domain of politics is that a progressive and radical political leader is usually followed by a conservative one and vice versa. They are both elected by the same political body, and actually during most of the cases, the exact same voters who now vote conservative voted for the liberal one in the previous election. This observation may express exactly this demand for a historical balance that is a mandatory condition for a society to operate efficiently. Actually, we may notice that the more conservative this leader is, the more progressive, radical and occasionally provocative his or her successor will be during most of the cases. We have witnessed this in the U.S. administration during the beginning of the twenty-first century. However, I think that it is more likely for the novelty-seeker to have voted for the conservative leader following an extended administration of the progressive one, than it is for the non-novelty-seeker that is probably more reluctant against the change. In that case, if we were able to examine DRD4 polymorphisms in voters who have voted for a different party in two consecutive elections, we might have seen that the ones who changed their minds and voted differently are primarily the novelty-seekers. In that case, those are the ones who actually determine the outcome of the elections, while the others, who are presumably the most rigid ones, represent a core body of political followers who are, in general, more reluctant to change political ideology. Of course, such an experiment has not been done as far as I know, but it might have provided quite informative insights regarding the contribution of genetics against political choices.

Vice versa, in the hypothetical societies that are practically devoid to the novelty-seeking 7-repeat allele, or analogous alleles that can substitute for, and also increase or stimulate the novelty-seeking behavior, they are quite likely to be condemned to static. While we can imagine that these societies will have a great collective perception and will be able to establish big and stable societies, they would still need their explorers and novelty-seekers to expand and liberals to doubt given perceptions and drive political advancement.

CHAPTER 8

Serotonin Transporter and the Emergence of Collectivism

The emergence of collectivism, as opposed to individualism, represents a type of behavior that signifies important differences between the way that Eastern and Western societies have been organized. Previously, some hints regarding the onset of collectivistic behavior have been described in view of the polymorphisms in the DRD4 gene and the specific behavioral patterns they were associated with. However, according to the results of several genetic analyses, another genetic locus displays a stronger and more direct relation to the development of a collectivistic behavior. It is related to the activity of serotonin.

Serotonin is an important neurotransmitter of the central nervous system, regulating many psychological traits and behaviors. It is released in the synaptic spaces between neurons, and its activity is terminated by a specific protein designated as serotonin transporter (5-HTT), which is responsible for its re-uptake from the pre-synaptic neurons. Several psychiatric drugs target this protein, underlining its significance in the regulation of conditions related to behavior. The regulatory region in this gene is not the same in all individuals, but differs in its length due to the different repetition number of a core sequence. Thus, some individuals have longer versions of this polymorphic gene while others have shorter versions.

Two major alleles and several minor ones have been identified in the population. The so called major, most common allele has been designated as the *s* (short) allele, consisting of 14-repeats, and the *l* (long) allele, which has 16 repeats. These variants of the serotonin transporter gene display differences in their activity: The shorter allele is less active than the longer one, and thus serotonin is re-introduced less efficiently in the neurons. As a consequence, the neurohormone displays prolonged activity, since it stays for longer periods outside the cell, at sites where it can be active and capable of eliciting biochemical signals. Related to this finding is the observation that, according to the results of many studies, the presence of the *s* allele has been linked with neuropsychiatric disorders, while the *l* allele is linked with better response to antidepressants. The question now is how all these are interpreted in the context of physiological responses and personality traits, within the borders of what we perceive as "normal".

While both *s* and *l* alleles are present in virtually all populations and ethnic groups examined so far, their frequency in the different groups is striking in terms of their geographic distribution. Specifically, the *s* allele was found to be much more common in the populations from the Far East, at frequency that range around 75%, as compared to the Caucasians at which the frequency ranges around 40%. Table 3, which has been compiled from different independent studies, shows these differences between various ethnic groups (Murakami et al., 1999; Goldman et al., 2010). Besides the reported differences in the frequency of the *s* and *l* alleles between Easterners and Westerners, one can see that Native Americans have an *s* allele frequency of around 65%, which is slightly lower but comparable to that of the people from Asia (it ranges between 67% for the Taiwanese and 80% for the Japanese). This is probably also due to the origin of the native Americans from the original migration of Asian populations towards this continent. In addition, Africans have very low frequency of the *s* allele, around 27% which is lower from that of all other populations tested.

Table 3. Allelic frequencies of the *s* and *l* polymorphisms in 5-HTTLPR in various ethnic populations (adapted from Goldman et al., *Depress Anxiety. 2010* March; 27(3): 260–269, 2010)

Ethnic group	% s allele*	% l allele*
Caucasians	43	57
Taiwan	67	28
Chinese	72	26
Japanese	80	19
Korean	71	29
Africans and African Americans	27	72
Native Americans	65	35

*The sum of *s* and *l* alleles is lower than 100% in many cases, because rare alleles, different than those two have not been included in the analysis.

8.1. Association between 5-HTTLPR and Depression

Various psychiatric conditions and disorders have been linked to the VNTR polymorphisms that exist in the promoter of 5-HTT. One such example is the association of the presence of one or two short alleles of the serotonin transporter and major depression. According to the results of such a study, individuals with the *ss* or *sl* genotype, when exposed to stressful life events, exhibited more depressive symptoms, diagnosable depression, and suicidality when compared to those bearing the *ll* genotype (Caspi et al., 2003). Subsequent meta-analyses, being studies that cumulatively analyze experimental results published in previous studies, confirm this association (Karg et al., 2011).

That the *s* allele, however, is only linked to pathologic behavior also acts as a sensitizer in the perception of depressive signals, as it does not reflect explicitly the whole picture. Taylor and co-workers (2006) studied not only how this allele affects the negative but also the positive signals. They found that if *s* allele-bearing people were exposed to positive signals, then the depression-related symptoms were less pronounced than the ones with the *l* alleles. Thus, the presence of the *s* allele sensitizes individuals not only against the pro-depression signals but also against anti-depression signals. In that case, we can view this allele as a multiplier of those signals, while the *l* allele operates as a "buffer" or a "filter": having it means that someone is less vulnerable against such stimuli. This may imply that the presence of the s allele may be essential in determining the tightness of social interactions, by influencing the dependency of the individuals for them.

Interestingly, this activity of the *s* allele is not extended to all types of positive or negative signals but only to those that are related to signals elicited by other individuals and reflect or elicit social support or relations. According to a study that evaluated the effects of a natural disaster, such as a hurricane (a negative signal that was not triggered by other people), the presence of the *s* allele had no difference. It did, however, when these hurricane victims received social support: If the support they received was not that good, then the *s* allele carriers were more likely to develop depression than the *l* allele carriers. When, however, they received good support, no considerable difference was found (Kilpatrick et al., 2007). Thus, the *s* allele appears to be able to interpret and modulate social responses if interaction with other people was involved. Consistent to that notion, individuals bearing the *s* allele are more receptive and sensitive to the social group they belong to or interact with, as compared to the *l*-allele-bearing individuals, who appear as less vulnerable against these stimuli. Due to that function, the *s* allele has been recognized as a *social sensitivity allele*. This hypothesis attributes an important role for HTTPLPR in the establishment and the maintenance of social bonds and the interdependency of the individuals, especially during certain forms of stress at which the requirement for social support becomes more prominent. In collectivistic cultures and societies in particular, social sensitivity is elevated, and this is imperative for the building of strong and tight social structures.

8.2. The Sociality Gene?

While the association of the *s* allele with increased social sensitivity, that is in turn related to the collectivistic behavior, is in tune with the major characteristics of Eastern cultures, the interpretation of these results is more complicated as regards the link between HTTPLPR and psychopathologies.

The prediction is clear: since the frequency of the *s* allele is higher in Asians and the *s* allele is linked to depression, then Asian people should exhibit a higher prevalence to the symptoms of depression than Westerners. Again, we have to face the same restrictions that refer to cross-cultural studies: Quantifying depression and related symptoms in individuals from different cultures, and especially when attempting to put together and make sense out of the results obtained from analyses of independent studies, presents several difficulties and limitations. These limitations range from the point of how depression is being diagnosed, to the ease by which individuals present themselves to appropriate practitioners in order to have their potential depression recorded, to their will in participating voluntarily is such exploratory study, and ultimately to the definition of a sharp borderline between what normal and what pathological is. Regardless of these restrictions, though, the answer is clear and probably intuitively anticipated: There is indeed a correlation that is present, but it is a negative correlation, actually. Populations with high frequency in the *s* allele, such as Easterners, have a lower instead of higher prevalence of depression than Westerners. In other words, people from Western cultures suffer more frequently from various manifestations of depression and general mood disorders than those of Eastern cultures (Chiao & Blizinsky, 2010). Reasons related to different criteria in classifying these conditions and reported incidences in different cultural environments, may, to some extent, explain these discrepancies, although not sufficiently.

We have to keep in mind that studies identifying the *s* allele as a depression-related allele involve, in principle, populations of the same culture. Thus, the specific effects of the cultural environment were excluded or diminished. Comparing results of studies that involved individuals of different cultural background could have solved this issue, but it is rather difficult to assume that exactly the same individual criteria have been applied in order to record depression effectively. No matter how strict and rigorous the criteria used were, a subjective factor could not have been avoided and ruled out.

Taking into consideration cumulatively the results of such analyses, another interesting association can be found between the frequency of the *s* allele and a specific characteristic, that at this time instead of referring to individual people, refers to whole cultures (Chiao & Blizinsky, 2010; Way et al., 2010). Such characteristic is related to the development of individualistic or collectivistic cultures, as described earlier in detail, when the former type of culture was defined as a predominant characteristic of ancient Greeks and their derivative Western civilization, while the latter was attributed to the Chinese and their derivative Eastern or Asian cultures.

We can easily see that in cultures with a rather individualistic perception of life, the frequency of the *s* allele is low, while the *s* allele is more common in people from collectivistic cultures. Figure 6 clearly shows this correlation. In this figure, data on the degree of collectivity in various different cultures were obtained by Hofstede (2001).

Various questions arise from these observations. First of all, coming back to the discrepancy described earlier, how is the high *s* allele's frequency accompanied by low prevalence of depressive symptoms and, generally, manifestations of anxiety and mood disorders in Asians? A proposed explanation is that individuals carrying the *s* allele, by developing a tendency for collectivistic values, at the same time provide themselves with a protective niche against symptoms of depression and also from stimuli that may trigger negative emotions and eventually psychopathologies (Chiao & Bizinsky, 2010). In other words, collectivism is protective for the ones in need and acts as a safety net that buffers the consequences of such negative stimuli. This is also in agreement with the previously mentioned role of the

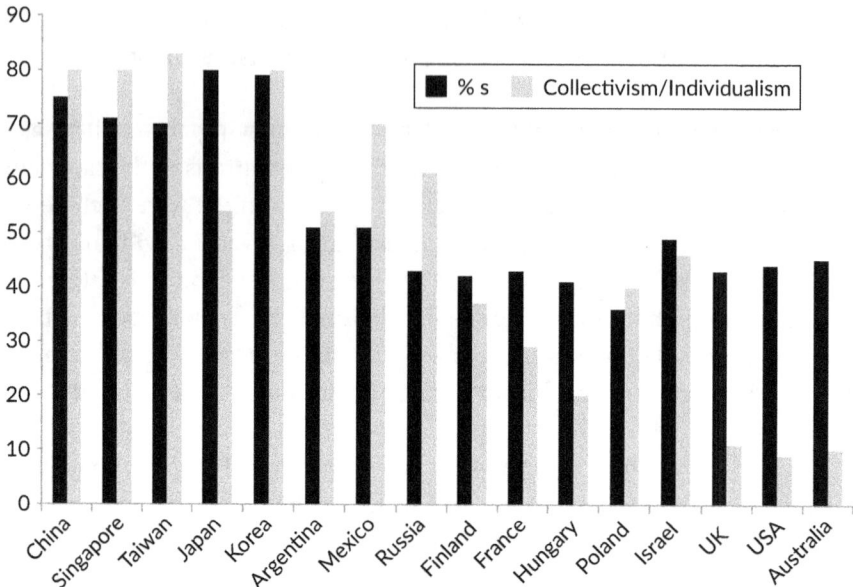

Figure 6. Correlation between the frequency of the s allele in 5-HTTLPR and collectivistic behavior. The data were adapted from Chiao & Blizinsky (2009) and Hofstede (2001).

s allele in interpreting social support. The opposite way of seeing things may also work. For example, the promotion of collectivistic tendency may precede the predisposition to depression. In that case, it is possible that individuals, by being more dependent to others they become more vulnerable to depression when they have to deal with adverse stimuli.

In any case, how did, at the first instance, Asian people exhibit higher s allele frequencies? Or, alternatively, did they have to develop (or be selected for this trait) this protective mechanism while Westerners did not? It is possible that this just reflects the s allele's frequency in the people who originally inhabited these areas.

It turns out, though, that the presence of the s allele offers another advantage that is also tightly related to the specific environmental conditions: epidemiological studies indicate that the collectivistic values offer an advantage against the transmission of disease-causing pathogens (e.g., typhus, malaria, tuberculosis), all of which exhibit high prevalence in geographical areas at which collectivistic values are more prominent (Chiao & Bizinsky, 2010; Fincher et al., 2008). So, people in these collectivistic societies are less vulnerable to the fatal effects of these contagious diseases. We also recently saw how efficiently China responded to the COVID-19 pandemic, a success that appears to be relevant to their collectivistic values and culture.

Selective pressure, though, represents just the one face of the coin because phenomena related to genetic drift can also offer satisfactory explanations as regards the high frequency of the *l* allele in Westerners when compared to the higher frequency of the *s* allele in Easterners. In that case, we have specific alleles enriched in different populations, which in turn trigger a series of traits that seem to function efficiently when combined with each other.

8.3. The Greek Confucius and the Chinese Aristotle

We can imagine the following scenario: First we consider the fact that the *s* allele predisposes for collectivistic behavior and strong social norms as a given. In that case, we can assume that originally, among the people who migrated towards Eastern Asia, the ones that carried the *s* allele had an advantage because the new environment, rich in contagious pathogens, favored individuals with the collectivistic behavior (or the populations-communities at which collectivistic behavior prevailed). The beneficial effects of the *s* allele, at the first instance, before structured societies were formed, were probably related to the contagious diseases present in the corresponding areas. However, other features of the *s* eventually started to exert themselves and prevail, and at this time, they were associated more, with the mode of social organization. Thus, they developed cultures and societies characterized by collectivistic values. In these cultures, the "whole" has a dominant role against the sum of its constituent entities (the individuals), and in order to function as such, the individuals that participate in this "whole" needed to exhibit high interdependency. In other words, the others, as a society, are needed to have a strong effect on each and every individual, establishing a spectrum of interrelationships, that operates as strong connective glue in these societies. In turn, among the various consequences in their respective collectivistic cultures, the high frequency of the *s* allele also protected the individuals against the psychopathologic symptoms of depression and related disorders. We can say, then, that it is Confucius and not Aristotle who was better suited for East Asia, because the *s* allele rendered the people that followed his teachings more successful in this particular environment. Besides the contagious diseases, we can also speculate that farming, due to the specific environment in the

East, could be practiced quite extensively and in order to do so, the people were required to establish a degree of collectivism. Vice versa, in ancient Greece, it was Aristotle who was better suited than Confucius, not necessarily because the *l* allele offered a particular advantage at first, but also because the *s* allele did not have a particular role to prevail. Assuming, however, that if among the people that originally inhabited ancient Greece, the frequency of the *l* allele was considerably lower than what it was, and thus, could not survive after a few (or several) generations, or that malaria was a common disease in these areas, then things might have been different and Confucius, or a Confucius-minded individual, might have been the philosopher with the greater impact on Western philosophy. In that case the culture that prevailed could have been a more collectivistic one. So, it was the specific environmental conditions that apparently drove the selection of specific alleles in certain populations, which in turn developed particular cultural attitudes and norms.

Someone may argue the operation of an alternative scenario that might actually have resulted in a similar result. That Easterners, despite their high frequency of the *s* allele, appear less vulnerable to its potential psychopathological consequences (assuming an overall lower prevalence of depression in the East than in the West), which may indicate that in a society of many more individuals with the *l* allele, the *s* allele appears to be disease-associated, just because it is reactive to the cultural majority dictated by the *l* allele-bearing people. This of course assumes that "depression" threshold is defined as such by matching closer what is instructed by the *l*-allele individuals. In a society that is dictated and characterized by the *s* allele's cultural norms, the *s* allele is not disease-associated and neither—of course—is the *l* allele. In that case, taking this line of thought a step further, one may argue that it is the *l* allele and the resulting cultural line that is the "disease-associated" one, while it is the *s* line that is the collectivistic one, and the line that is actually closer to people's individual needs and capacities.

The slight difference introduced by this latter scenario is that the disease- associated symptoms contributed by the *s* allele are rather reactive than causative per se. Indeed, individuals with East Asian origin who live in the U.S.A. suffer more frequently from depression than individuals living in East Asia, as evidenced by analyses at which the same criteria for the diagnosis of the disease were used (Chang 2002; Hwang & Myers, 2007; Way & Lieberman, 2010). Thus, in different (sub)-groups of people who carry the *s* allele at the same frequency (Asians), it is the *l* allele's frequency in the overall society that harbors these individuals that triggers (or protects them from) the psychopathologic behavior. In that case, what really matters is the combination of the *s* allele's frequency in the study group, as compared to that of the total *s* allele's frequency in the whole (super)-group in which the study group belongs to. This again brings into consideration the context, the environment as a significant determinant and essentially implies that what may classify an allele as a predisposing allele is not its actual presence in the experimental subjects but rather its absence from the others. Again, this underlines the significance of relative frequencies in alleles in order to conclude and extrapolate to their impact on society.

8.4. A Gene for Happiness

The results of behavioral genetic studies frequently attract headlines. They appeal to readers because they reveal hidden aspects of our personalities or maybe even because they rationalize certain behaviors, potentially providing alibis for tendencies we may have. Moreover, they justify specific social attitudes and reactions. This was also the case with a study by De Neve et al. (2010) that explored the genetic basis of happiness. Press heralded these findings as a step closer to a happier society, or for offering clues for more productive companies. The fact that these studies involve genetic analyses and specific experiments provides an increased amount of objectivity, at least as compared to analogous results obtained by more theoretical and classical approaches. Noteworthy, this particular research group included researchers who have published results on the genetic basis of liberal political ideology or leadership, as we discussed already or that we'll see shortly.

In the study about happiness, the investigators focused on the self-perception of happiness, asking participants the following simple question: "How satisfied are you with your life as a whole?" The potential answers ranged from "very dissatisfied," "dissatisfied," "neither satisfied nor dissatisfied," "satisfied," to "very satisfied." Subsequently, the results of these questionnaires, which involved about 2,500 individuals, were analyzed in combination with the allelic frequencies of 5-HTT polymorphisms. The experimental groups also involved twins, which is of particular importance in such studies because it permits the assessment of whether a particular characteristic possesses a genetic component, and if it does, to get an idea of the extent of this genetic contribution. Complex mathematical analysis followed, and the results confirmed that the perception of happiness bears a significant genetic constituent. More importantly, they provided specific evidence that bearing the *l* alleles of 5-HTT increases someone's chances to feel satisfied with his or her life. Interestingly, this does not apply simply to the presence or absence of the corresponding alleles (*ss* genotypes vs. *ls* and *ll* genotypes), but also extends to the number of the *l* alleles the individual bears: *ll* individuals have almost double the chance to feel very satisfied with their lives, as compared to the *sl* individuals. There is a dosage effect apparently.

That the *l* allele is more common in Caucasians, as compared to Asians, predicts that the Westerners should feel happier and more satisfied with their lives than the Easterners. It's practically impossible, though, to compare whole nations and cultures in terms of happiness. After all, each and every culture's purpose is to offer a "meaning" in peoples' lives, make them content, and ultimately, in its widest sense, to make them feel happy. Many parameters are involved in the quantification of such statements, and even the perception of happiness likely differs between cultures. On the other hand, even if the incidence of certain psychopathologies represents an indication of pathological unhappiness, again, for reasons discussed in detail elsewhere, no valid and accurate comparison can be made.

Intuitively speaking, though, I have to admit that I would rather expect Asians to be happier in general, than Westerners. I cannot support this by specific arguments, but I think that the reason for that is related to the individualistic approach of life that the people possess in Western societies: By operating under such individualistic norms, it is unavoidably stressful, a condition that likely operates at the expense of the perception of individuals' happiness. In the absence of the protective niche that is present in collectivistic societies—and as previously mentioned likely protects from the symptoms of depression and of other psychopathologies—the perception of happiness should be reduced. We can witness this in our individual lives, that insecurity is tightly related to unhappiness. The high frequency of the so-called happiness-associated allele in Westerners suggests the opposite: People from the East should feel less happy than people from the West.

I can only think of two explanations that may argue against this phenomenological discrepancy. First, evaluating, and even more importantly, self-assessing the level of satisfaction in someone's life, hides a latent but definite individualistic notion: Only under individualistic terms is the satisfaction in an individual's life central and intrinsically related to the notion of happiness. Consistent with a collectivistic notion for life, this attitude of pursuing happiness is probably replaced by other notions, at which the Westerners' perception of happiness is of reduced significance. In simple words, what makes Asians happy and satisfied is different from what makes Westerners happy. Even more important than that is the fact that while Westerners seek happiness at the individual level (and eventually reports it on an individual basis), Easterners are likely to attempt to pursue happiness as part of the greater whole in which they belong. After all, the pursuit of happiness is not a major issue for Easterners, for whom a more appropriate notion is probably contentment. Thus, a comparison cannot be made, and even the notion of satisfaction is meaningless. It's like the oranges and apples comparison.

Another possibility is that, despite the restrictions and limitations related to the stress and tension of their individualistic society, Westerners feel happier than Asians because they view their lives as a trip with a specific starting point and a well-defined destination. This notion opposes the circular mode of life that is closer to the Asian perception. Thus, Westerners are able to set and achieve specific and measurable goals that can be perceived as an achievable milestone for satisfaction and happiness. The same question in the Asians might be irrelevant. In other words, for Westerners, it's the trip instead of the destination that may make someone happy.

This notion has been successfully expressed by Konstantinos Kavafis, the famous early-twentieth-century Greek poet who started his poem "Ithaca" as follows:

> When you set out on your journey to Ithaca,
> pray that the road is long,
> full of adventure, full of knowledge.

8.5. Combined Polymorphisms in DRD4 and 5-HTT Promoter

The two polymorphisms discussed earlier, namely the 7-repeat allele for DRD4 and the *l* allele for 5-HTT, have been associated with Western-type behaviors. The first one, among the other traits for which associations were found, was linked to novelty-seeking, liberal behavior, risk-taking, and substance abuse, while the latter with more individualistic behavior. It appears that these traits are complementary in establishing the personality of Westerns. Novelty seekers and inventors need to make decisions and take risks by operating on a more individualistic basis than those prone to a more collectivistic behavior. Since none of these two polymorphisms characterize exclusively the two populations, Asians and Westerners, and the two genes are not linked genetically, many individuals in both geographical regions should as they indeed do have the 7-repeat of DRD4 and the s allele of 5-HTT, or the 4-repeat DRD4 allele with the s allele of 5-HTT. An anticipated question, of course, in that case is what happens if these two polymorphisms co-exist in the same individuals? Do novel behavior patterns arise, or are the pre-existing ones becoming more intense? In other words, does the presence of both 7-repeat and l alleles indicate a genetic predisposition for a super-typical Western-type behavior and vice versa? Does the presence of the s allele for 5-HTTPLPR, combined with the absence of the 7-repeat allele for DRD4, imply a pattern of behavior that is very typical for Easterners?

Fortunately (or un-fortunately!), things are more complex than that. Let's see what such analyses, that co-examined the presence of both alleles, found. In fact, several studies are available in which the combination of 5-HTT and DRD4 was explored. However, their vast majority concentrated in their effects on certain psychopathological conditions. We'll try here to omit these studies and focus only on some of them that have limited their analysis at certain behaviors but not pathologies. In 2004, a study by Szekely et al. (2004) reported that at least in Caucasians, an s/s 5-HTTLPR combined with the 7-repeat DRD4 genotype showed a higher mean harm avoidance[21] score (Stallings et al., 1996) than the other groups. This trait is linked to behaviors such as pessimism and fearfulness. Noteworthy, while the corresponding DRD4 genotype is the one that is common to Westerners, the 5-HTTLPR genotype is the one that corresponds to the alleles that are more prevalent to Asians.

In a different experimental setting, involving evaluation of temperament in infants by using a set of standard episodes eliciting fear, anger, pleasure, interest, and activity, Auerbach et al. (2001) reported that a certain combination of DRD4 and 5-HTTLPR alleles was associated with a shorter duration of looking during block play. These genotypes again were the long, novelty-seeking alleles for DRD4, combined with the s alleles for 5-HTTLPR. At the same trend were also the results of another study that took place in 15-year-old adolescents: individuals carrying two copies of the 5-HTTLPR short allele (s allele) and the DRD4 7-repeat variant scored highest on aggressive and/or delinquent behavior compared to other genotypes (Hohmann et al., 2009). In another study performed by Lakatos and co-workers in 2003, infants with the 7-repeat DRD4 allele that were also homozygous for the short form of 5-HTTLPR (*s/s*) showed more anxiety and resistance to the stranger's initiation of interaction than infants bearing the 7-repeat DRD4 allele along with the *l* allele for DRD4.

Contrary to this "bilateral" association between the "Western" DRD4 allele and the "Eastern" 5-HTTLPR allele, Auerbach et al. (1999) reported that infants with the *s/s* 5-HTTLPR polymorphism who were also lacking the long DRD4 alleles associated with novelty-seeking showed more negative emotionality and more distress to daily situations, temperament traits that are perhaps the underpinning of adult neuroticism. This study took place in two-month-old infants.

With the exception of this last study, which showed that the "Eastern" type alleles (*s* for 5-HTTLPR and the short (non-7-repeat) DRD4) exhibited an interaction, the other two studies showed that if an interaction is there, it is between an "Eastern" and a "Western" allele. In that case, and depending on the exact experimental setting, this interaction was exemplified as a shorter duration of looking during block play in infants, evidence for aggressive behavior in adolescents or harm avoidance in adults.

These results are far from being definitive, of course, and probably discussing the implications of these findings is premature. What is important is to notice that if, indeed, certain alleles were not only associated but actually contributed to a Western- or Eastern-type behavior, this doesn't necessarily mean that by putting them together we will obtain for sure a kind of interaction or synergy, as no such data were recorded.

Interestingly, though, and if we indeed have obtained an interaction with specific combinations (one Western- and one Eastern-type), as these preliminary studies proposed, an interesting argument could be made: All these traits that are influenced by such interactions can be considered as negative (not pathological, though) in the long run, if we view them at the level of the society: Shorter duration of looking at the infants can be indicative of less-efficient learning capabilities, while increased harm avoidance and "fear" of strangers can eventually reduce personal interactions and develop eventually as an obstacle in the overall advancement at the level of society. Tendency for aggression in adolescents is also an apparent negative trait at the level of society.

So, if these combinations combinations between "Eastern" and "Western" genes indeed exhibit some "disadvantage," this might be indicative for the operation of a latent mechanism that, besides all other predominant mechanisms discussed earlier (genetic drift during initial migrations and natural selection and "fit"), has augmented the diversification between Eastern and Western people: Differing in two instead of a single trait makes individual people and populations more distinct.

8.6. MAAO and Collectivism

Among the genes that potentially affected the development of collectivistic behavior is the one encoding for the enzyme monoamine oxidase A (MAOA) that breaks down neurochemicals such as serotonin and dopamine. This gene is also polymorphic at its regulatory region: Different alleles exist in the natural population that differ in the number of repetitions of a core tandem repeat (upstream element variable tandem repeat, uVNTR). Among these alleles, the ones with three and four repeats are the most common in the global population, with the latter exhibiting higher activity than the former (Pai et al., 2007).

As shown in Table 4, interesting differences also exist in the frequencies of MAOA-uVNTR in the world population. In general, results from several studies indicate that the 4-repeat allele (individualistic) is more common than the 3-repeat (collectivistic) allele in Westerners, as compared to East Asian people. While the ratio between the 3-repeat and the 4-repeat alleles in Easterners is in the range of 65% vs. 35%, for Westerners this ratio is reversed to 35% vs. 65% (Pai et al., 2007; Jorm et al., 2000; Deckert et al., 1999; Hamilton et al., 2000).

Certain behavioral traits have been linked to the polymorphisms of this gene. The MAOA-uVNTR has been associated with differential sensitivity

Table 4. Allelic frequencies of the 3 and 4-repeat polymorphism in MAOA-uVNTR in various ethnic populations (data were adapted from Pai et al., 2007, *Forensic Science Journal 2007*; 6(2): 37–43)

Ethnic group	% 3-repeats*	% 4-repeats*
Chinese Han in Taiwan	63	36
Japanese	62	38
Asian/Pacific islanders	61	38
White/non Hispanic	33	65

*The sum of s and l alleles is lower than 100% in many cases because rare alleles, different from those two, have not been included in the analysis.

against the social environment in a manner that is consistent with that among the individuals with the low-expression allele (the 3-repeat allele, that is actually more frequent to Easterners), who had adverse childhood experiences; they also had the lowest levels of antisocial or violent behavior (Kim-Cohen et al., 2006; Widom et al., 2006; Ducci et al., 2007). The explanation for the aforementioned tendency and the link to the collectivistic one, as opposed to the individualistic behavior, is likely similar to that described earlier for 5-HTTLPR. It is tightly linked to the existence of a strong social support network that eventually strengthen societal bonds and ultimately collectivistic behavior and norms. Again, collectivistic behavior appears to protect from negative stimuli.

Also analogous are the findings for another polymorphism in the MAOA gene, at position 118, at which there is an A to G substitution that, in turn, results in the A and the G allele, with the latter being considered the "social sensitivity allele" that is most common to Easterners (Way & Lieberman, 2010). However, among maltreated children, the ones with high expression MAOA alleles were less likely to develop antisocial problems, implying that the MAOA gene is an important modulator of the sensitivity and of the subsequent responses of certain environmental stimuli (Caspi et al., 2002).

In addition to the interpretation of certain stimuli, an "endogenous" pattern in specific negative behaviors exists, which is not necessarily linked tightly to the exposure at such negative stimuli. Indeed, a direct association between the MAOA-uVNTR polymorphism and aggression has also been proposed: Individuals carrying low-expression versions of the MAOA-uVNTR alleles (such as the 3-repeat allele) exhibited higher trait aggression than the individuals with the high-expression alleles (such as the 4-repeat alleles), according to a study that involved a sample group consisting of about 28% European-American, 40% Asian, 15% Hispanic, 6% African-American, and 9% "mixed" or other, and was performed at the University of California at Los Angeles community (Eisenberger et al., 2007). This study, notwithstanding the statistical significance of the results, involved only a limited number of participants (n=32) and was largely based on a self-assessment of aggression by answering questions such as "How bothered do you feel about … having urges to beat, injure, or harm someone?" "… feeling easily annoyed or irritated?" Another potential limitation can also be that it was performed in the "Westerners'" setting in which 3-repeat individuals were operating in a 7-repeat context (the majority allele in the given society). In that case, the lack of social support

103

that is conferred by the 3-repeats allele is missing, and probably aggression in the individuals carrying the 3-repeats allele is enhanced.

Also analogous were the results of another experiment synthesizing approaches of psychology and behavioral economics (McDermott et al., 2009). In this study, the investigators found that the low-expressing MAOA alleles predict aggression following provocation. The experimental setting was based on the premise that individuals were making a payment to punish those they believed had taken money from them by administering varying amounts of unpleasantly hot (spicy) sauce to their opponent (McDermott et al., 2009). Thus, the investigators were able to study the effects of the specific MAOA genotypes in association with the willingness of the subjects to engage in physical aggression toward another when they believed that they had money taken from them. Their analyses indicated that low-activity MAOA individuals exhibited increased chances to engage into this aggressive behavior. Interestingly, the hot sauce they used for the punishment could have been traded for money. Therefore, it had a specific monetary value. So, the "satisfaction" of administering this punishment comes with a certain cost to the individuals that render it. In that sense, their behavior is not costly only to the others (the ones they want to punish) but also to themselves! This behavior is termed "spite" and is thought to represent the "neglected ugly sister of altruism" (McDermott et al., 2009), exemplifying how close altruism can be with this type of aggressive behavior.

In addition to those studies, another experimental study in male mice that do not have a functional MAOA gene showed that these animals are more aggressive than their wild-type counterparts, which fully supports the notion that variation in MAOA expression is causatively linked to the development of an aggressive behavior (Cases et al., 1995).

It is finally noted that a controversial study has also linked MAOA polymorphisms to the acquisition of a warrior behavioral pattern, according to data obtained after analysis of MAOA polymorphisms in Maoris in New Zealand. The corresponding study has attracted several headlines, as it has identified a direct genetic predisposition to aggression (Lea & Chambers, 2007).

Given the distribution of the corresponding polymorphisms, the aforementioned findings predict that aggression should be more common to Easterners than Westerners. While again a formal comparison cannot be made, it appears that such a statement contradicts the anticipated observation, that aggression should be reduced in the collectivistic societies that are structured in a manner at which the seeking of harmony plays a major role.

However, taking together all these results about MAOA, we may hypothesize that while the low-expression alleles are linked to aggressiveness, the "anti-social" effects of this apparently negative behavior are minimized, and likely masked, by its concomitant association with the collectivistic behavior which is eventually promoted. The latter trend apparently functions protectively, by scaling down the probable negative consequences or even the expression of aggressive behavior in society. This can be the case with Easterners, that while they possess the 3-repeat, low-expression allele at high frequency, aggressiveness does not apparently predominate within their societies, because other protective norms have been established, such as those associated with collectivism. Although more individuals may be more prone to aggression in these societies, the concomitant collective norms that they develop are protective. In other words, individuals feel that they may have more to lose than to gain by being aggressive. Such modes of action not only protect the individuals and the societies from the adverse effects of aggression, but also stabilize the collectivism.

In addition to that, the fact that MAOA-associated aggression (low-expression alleles) is particularly linked to provocation (McDermott et al., 2009) may imply additional clues regarding the function (and maintenance) of such collectivistic behavior. It may indicate the operation of strong societal norms that may nurture the concepts of revenge, punishment, and maintenance of strongly hierarchical societies. The latter can be seen in the East, in which certain codes of honor transcended the history of the Far East and played their role in the maintenance of the rigidness in corresponding societies. If the fear of punishment is strong and imminent, this operates at the expense of individualism. Social structures are rigid, facts that may be expressed as a tendency to collectivistic behaviors.

In Western societies, in which the protective effects of collectivism are not present, aggressive behavior is more easily expressed (and people get away with it since punishment is not that strong and imminent), and occasionally it even becomes dominant, although at the genetic level it may be initiated by a smaller number of individuals. At this point, of course, we have to recall some thoughts we described in chapter 6, on discussing the individuals' as opposed to the populations' traits that may explain, at least in part, why a higher number of aggressive individuals does not necessarily mean "more aggressive" populations and eventually societies.

CHAPTER 9

COMT, Altruism, and the Evolution of the "Warrior versus Worrier" Strategies

Catechol-O-methyltransferase (COMT) is an enzyme that is responsible for the inactivation of catecholamines in the synaptic cleft. Widely known and extensively studied, catecholamines are the hormones dopamine, adrenaline, and nor-adrenaline. Among their other functions, they regulate the response against stress that, in turn, in vertebrates, is exemplified by a "fight or flight" reaction against threats (Cannon, 1929). This response simply means that when an animal (including humans) encounters a specific threat, immediately it has to select between dealing with the threat (fight) or depart in order to avoid it (flight). In any case, the ultimate purpose of this response is to deal successfully with the danger and, among its various rapid adaptations, also involves certain changes in the body's physiology that aim to give a burst of energy and strength. Whether it's going to be "fight" or "flight" depends on various parameters, but at the level of the body's physiology, both trigger a set of similar neuro-chemical adaptations.

Catecholamines play a major regulatory role in this process and the development of the corresponding response. The fact that COMT inactivates the catecholamines predicts that this enzyme can be a modulator of this response in a manner that is consistent with that under stress, increased COMT activity results in lower dopamine levels. This response in turn is associated with improved dopaminergic transmission and better performance. More efficient dopamine clearance can be seen as a better acute responsiveness against the specific stimulus. Indeed, a polymorphism in the COMT gene exists in the population and is directly related to the activity of COMT. At position 158 of the protein, a G to A transition exists in the population that changes the aminoacid valine (Val) to methionine (Met). This change is related to higher activity of the Val-bearing allele, as compared to the allele bearing Met. Thus, Val-bearing individuals, under stress, exhibit better performance, as compared to the individuals with the Met allele.

Now, let's see how these neurophysiological changes might be linked to behavioral patterns and strategies. Intrinsically related to the final interpretation of the "fight or flight" response are also the "warrior or worrier" strategies. Roughly speaking, a "warrior" is someone who exhibits an advantage in the processing of aversive stimuli and who performs better under stress; a warrior is characterized as one who achieves maximal performance despite threat and pain and who is more efficient in dealing with the threatening environment. Simply speaking, this individual is more intuitive and can be described as a person of action.

On the other hand, the "worrier" is someone that has an advantage in memory and attention tasks, is more exploratory and efficient in complex environments, but who exhibits worse performance under stressful conditions (Stein et al., 2006). COMT polymorphisms have been associated with this behavior, with Val-allele individuals being the warriors and *Met*-allele individuals the worriers.

In view of these findings, let's see how the two alleles of COMT are being distributed around the world according to Palmatier et al. (1999). As shown in Table 5, the G allele is more common to Asian populations, as compared to Europeans.

While in Europeans the G allele's frequency ranged at about 40%, in East Asian populations its frequency is above 70%. So, it appears that Asians are likely to be warriors while Westerners worriers. According to the results of Palmatier et al. (1999), South American populations had high but variable allelic frequencies for the G allele, being 0.66, 0.81, and 0.99 for the native Surui, Karitiana, and Ticuna[22] people, respectively. A similar observation was made for the African populations tested. For Africans, this variability is not surprising, since Africa, is the source of all polymorphisms found in present populations. For the South American populations, it may simply reflect different allelic frequencies in the original populations that settled these areas and established the corresponding populations. It also reflects the genetic similarity of the South Americans and the East Asians.

Table 5. Allelic frequencies of the G and A alleles of COMT in different world populations. (Data adapted from Palmatier et al., 1999, *Biol Psychiatry 1999, 46, 557–567*)

Ethnic group	G allele frequency	A allele frequency
Africa		
Biaca	0.93	0.07
Mbuti	0.78	0.22
Yoruba	0.63	0.38
Europe		
Irish	0.48	0.53
Danes	0.38	0.62
Finns	0.47	0.53
Mixed Europeans	0.44	0.56
East Asia		
Cambodians	0.71	0.29
Chinese	0.69	0.31
Japanese	0.79	0.21
Atayal	0.85	0.15
South America		
Ticuna	0.81	0.19
Surui	0.66	0.34
Karitiana	0.99	0.01

Similar were the results obtained from the HapMap project and its sequel, the 1000 genomes project.[23] Table 6 shows the frequencies of the A and G allele according to the 1000 genomes project for the single-nucleotide polymorphism designated as *rs4680,* which corresponds to the COMT polymorphisms discussed above. In this table, Western populations include Utah residents with European ancestry and Italians from Tuscany.

Of course, the same argument can also be made for European populations, since the European continent was settled by successive waves of migration events, but in that case, COMT allelic frequencies for these populations are very similar. Why this variability persisted in South Americans is hard to speculate. It may reflect the extent of interbreeding between these people, a possibility that can be tested by analyzing allelic

Table 6. Frequencies of rs4680 polymorphism (G/A) in the COMT gene in different populations according to the HapMap project

Population	G allele frequency	A allele frequency
African ancestry in Southwest USA	0.730	0.270
Utah residents with Northern and Western European ancestry from the CEPH collection	0.535	0.465
Han Chinese in Beijing, China	0.684	0.316
Gujarati Indians in Houston, Texas	0.563	0.487
Japanese in Tokyo, Japan	0.716	0.284
Luhya in Webuye, Kenya	0.712	0.288
Mexican ancestry in Los Angeles, California	0.602	0.398
Mende in Sierra Leone	0.765	0.235
Tuscan in Italy	0.547	0.453
Yoruban in Ibadan, Nigeria	0.694	0.306

frequencies in other polymorphic genetic loci. It may also be associated with specific advantages conferred by the corresponding alleles to these populations. We also need to keep in mind that the aforementioned populations in South America were very small, and thus, the consequences of random drift phenomena related to the genetic constitution of the founders might have been exaggerated.

An interesting observation regarding the evolution of the COMT alleles is that in non-human primates, such as gorillas, chimpanzees, bonobos, and orangutans, only the G allele that corresponds to Val (warrior allele) has been detected (Palmatier et al., 1999). So, the earlier human ancestors had the Val allele. This implies that the A allele, which corresponds to the aminoacid Met (worrier allele), appeared more recently in evolution, after the divergence of the ancestors of modern-day humans. So, simply speaking, our evolutionary ancestors were originally warriors and then they become worriers!

By using a financial-based experimental setting, Reuter et al. (2010) reached another interesting conclusion regarding the possible contribution of the COMT polymorphisms in regulating altruistic behavior. According to a biological definition of altruism, this behavior results in the reduction of the individual fitness but increases the fitness of other individuals in

110

the population. Thus, it is tightly associated with social behavior. And in order to operate as such it is required to maintain some balance between the G and A allele that should coexist in the population. In this study, the investigators evaluated the amount of money donated by an individual to a poor child in a developing country after the subject earned money by participating in two straining computer experiments (Reuter et al., 2010). In this study, the participants, after having earned money in a gambling experiment, had the chance to donate a fraction or all of their money to a cute little girl from Peru. Analysis of the amount of money donated (altruistic tendency in behavior) in association with the corresponding COMT genotype showed that individuals carrying at least one Val allele donated about twice as much money, as compared to those participants without the Val allele. As expected, this research attracted several headlines, since the news for the identification of an altruism gene may have various implications in many aspects of life. The fact that altruistic tendencies possess an element of impulsivity indicates that it is not surprising that altruism is associated with the behavioral pattern of warriors (Easterners). On the other hand, the worriers, who are not that impulsive and often attempt to predict and consider the long-term consequences of their actions before they act, are more reluctant to behave as altruists, since this may cost them in the long run (the Westerners). Moreover, the latter's altruistic behavior is more likely exerted only when the associated benefits of altruism are proximal. In a recent study that examined the tendency toward altruistic behavior in a group of Americans, as compared to a group of Chinese, the former appeared more altruistic than the latter. When the investigators assessed the willingness to be altruistic towards outgroup members whose intentions were uncertain, the Chinese exhibited higher levels of altruism than the Americans (Lee et al., 2008). So, in the Americans, a hidden element of reciprocity is latent in their action of altruism.

Let's summarize all these findings: Of the two COMT alleles, the ancestral one that corresponds to the amino-acid Val is associated with the "warrior" behavior, an altruistic tendency, and is more common to Easterners. On the other hand, the allele corresponding to the amino acid Met is more recent in terms of evolution, is associated with "worrier" behavior, exhibits less intense altruistic behavior, and is more common to Westerners. Do all these make sense in terms of the current Eastern and Western cultures as we perceive them? As regards the altruistic tendency and by keeping in mind all these factors we described above regarding individualism and collectivism, Westerners are probably less altruistic than Easterners; this is in line with their individualistic norms. Furthermore, the Easterners that, according to this analysis, are the altruists are the ones that have such frequencies in the HTTLRP and MAOA alleles that instruct the establishment of a collectivistic society. In turn, this efficiently provides social support to the individuals that need it.

Therefore, at the level of society, it emerges that the specific Met-bearing COMT allele contributes to the buildup of Western individualism. Opposed to this, Easterners' increased frequency of the Val-bearing "altruistic" allele fits quite well with the construction of a collectivistic society: You have to be an altruist at some degree in order to understand the benefits of collectivism. By being a pure individualist, you only understand "good" as defined and reflected by your sole existence.

It is probably more complicated than that as regards the warrior and worrier attitudes. Warrior-like behavior is the one that tends towards actions dictated by impulsivity and confrontation and eventually may result in a constant turmoil. These notions are by far more suitable to describe Westerners, as compared to Easterners. Life in a Western and Westernized society is competitive and stressful, with a diminished sense of security, since nothing should be considered as a given and ultimately everybody has to fight in order to gain and then maintain his or her position in the society. However, because Westerners genetically are more prone to the "worrier's" behavior, they appear to prefer compromise over confrontation, stoicism, and the resolution of the disputes in a longer-run perspective than guided by a short-term outcome. An apparent contradiction develops, especially as regards the COMT-predicted preference of Westerners for compromise and stoicism (that appear as collectivism-related attitudes). A simple explanation for this discrepancy from the anticipated, "logical" observations and predictions could be that the contribution of COMT variability as regards this particular aspect of behavior is "recessive," as compared to that of other genes, such as the HTTLPR and MAOA described in previous chapters. In other words, maybe in Easterners, the "warrior"-prone COMT allele encoding for Val, contributes less to collectivism than the Met-bearing "worrier" allele, but its effects are surpassed by the high prevalence of the s allele for 5-HTTPLPR.

Another plausible explanation is that what signifies the aforementioned discrepancy should not be the high incidence of the "warrior's" Val allele of Easterners, but rather the higher prevalence of the "worrier's"

Met allele in Westerners. Consistent with this notion, Westerners have built quite compound societies characterized by a complicated structure reflected in all aspects of life. Eventually and unavoidably, problems that occasionally may appear should seek for a long-term resolution that could not be solved by an individualized approach. On the other hand, people that exhibit increased memory and attention, as well as efficiency under complex environments, would be better suited for these complex societies, such as those that characterize Western culture. To that end, a "worrier's" behavior appears to be more suitable than that of the "warrior's" in the West. In that case, it is the feature of the increased efficiency in complex conditions that supersedes the altruistic behavior that in turn characterizes the contribution of the COMT allele in Westerners. That COMT-conferred altruism does not become obvious in the West is probably due to the fact that the contribution of the corresponding allele is masked by other, "non-altruistic" alleles that are present in these populations and signify a "non-altruistic" behavior.

Of course, even for that argument—namely that altruism in the West is not prevalent—several counterarguments can be made. For example, the whole notion of giving to the ones in demand, a concept strongly advocated by Christianity, is intrinsically associated with the advanced societies of the West. Contemporary examples are the widespread existence of various charity organizations, and even the existence of several campaigns targeting the disadvantaged populations of today's world. Voluntary giving constitutes an integral part of Western culture. But again, these can all be notions developed to diminish apparent gaps in the "connective glue" of the society that follows individualism. The concept of "charity," including that advocated by Christians, has been harshly criticized in the Western world by various lines of political and philosophical thought.

Besides this interesting link between the COMT gene and the onset of behavioral patterns, such as the "warrior vs. worrier" strategy, polymorphisms in this gene have also been associated with certain psychopathological and neurological conditions. These include Parkinson's disease (PD), at which the low-activity allele (A) was associated with increased risk for PD (Kunugi et al., 1997) and others, such as homicidal behavior in schizophrenic patients (Kotler et al., 1999) or obsessive-compulsive disorder (Karayiorgou et al., 1997). These are all pathological behaviors and conditions. Interestingly, in the latter study the corresponding association was more pronounced in the males. Noteworthy, in all cases the allele

113

associated with pathology was the low-activity allele (A) that encodes for the amino acid Met. This is also the allele that is related to the worrier behavior, and is more prevalent to Westerners.

So, it appears that the benefits conferred by the Met allele that were related to the increased ability of individuals to function efficiently under complex conditions were compromised to some degree by a sensitization against certain pathologies. This reminds the "no good without bad" dualistic principle that is seen in various philosophies, and which range from the Yin-Yang Taoist principle to the teachings of Augustine, who claimed, "Nothing evil exists in itself, but only as an evil aspect of some actual entity."

In the case of COMT a possibility is that when the structure of social organization increased in complexity following the individualistic cultural line, individuals with the Met polymorphism had an advantage that was associated with the benefits conferred by the development of a "worrier's" behavior. An unavoidable consequence, though, was the predisposition to certain pathologies, such as PD and obsessive-compulsive disorder.

9.1. COMT and Taxes

How all these associations work when we compare them in different populations instead of within a single population? Does the frequency of COMT alleles correlate with any personality traits that can be evaluated on the society level? To do that, we would require quantifiable characteristics that relate to some information regarding the corresponding populations as a whole and differ in different communities. Financial indices on taxation, for example, may work for this since they correspond to quantitative measures that may reflect cultural traits. Furthermore, taxation, despite the limitations, describes aspects of the society and how this is organized. The World Bank provides such data for all countries in the World;[24] this allows comparisons of various countries with the frequency of various polymorphic alleles in the corresponding populations. Such comparison in 29 different countries (Colombia, Mexico, Peru, Puerto Rico, Brazil, Finland, UK, Spain, Italy, Germany, Greece, France, Sweden, Egypt, Denmark, China, Japan, Vietnam, Thailand, Korea, Bangladesh, India, Pakistan, Sri Lanka, Barbados, Nigeria, Kenya, Siera Leone, Gambia) was attempted for the variations in COMT gene and provided a series of interesting associations. It doesn't matter where each specific country is

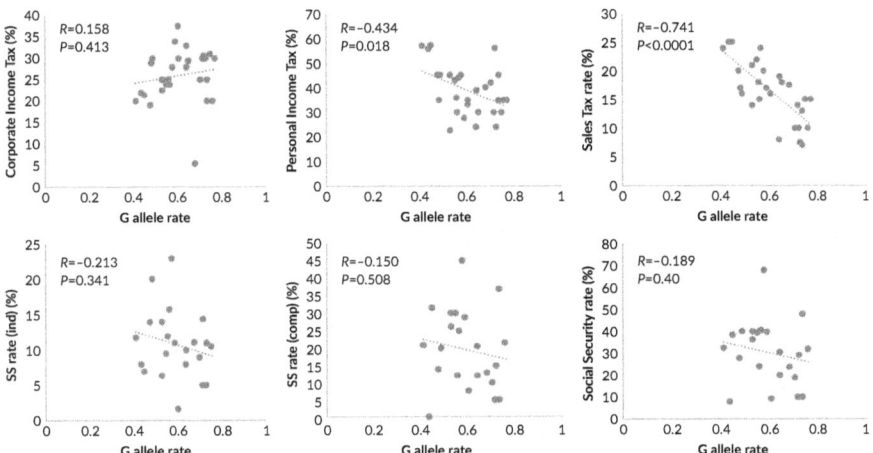

Figure 7. Scatter chart plot between the incidence of the G allele of COMT in different populations and a series of financial indices related to taxation in the corresponding countries. (R and P values, Pearson correlation) are indicated.

but rather the overall trends that are emerging. With regards to taxation, this analysis showed that countries with populations having higher rates of the G allele incurred lower sales tax, a highly significant correlation (R=−0.741, P<0.0001) (Figure 7). In this graph, each dot corresponds to a different country.

A similar trend was also found between the prevalence of the G allele and personal income tax. The correlation was again significant but weaker (R=−0.434, P=0.018). With regards to various indices reflecting economic growth, a negative correlation was revealed between the G-allele's prevalence and the national income (R=−0.609, P=0.0008), as well as the adjusted net savings per capita (R=−0.490, P=0.009). A positive correlation was found between G allele's prevalence and the adjusted net national income (annual % growth) (R= 0.412, P=0.032) (Figure 8). No significant association was found between the rates of G allele and each of adjusted net national income (current US$) and adjusted net national income per capita (annual % growth) (Figure 8).

A similar analysis was also performed by using various indices reflecting sustainability which showed a positive correlation between the rate of G allele and natural resources depletion (% of GNI) (R=0.396, P=0.037) and a negative correlation with educational expenditure (R=−0.42, P=0.024)

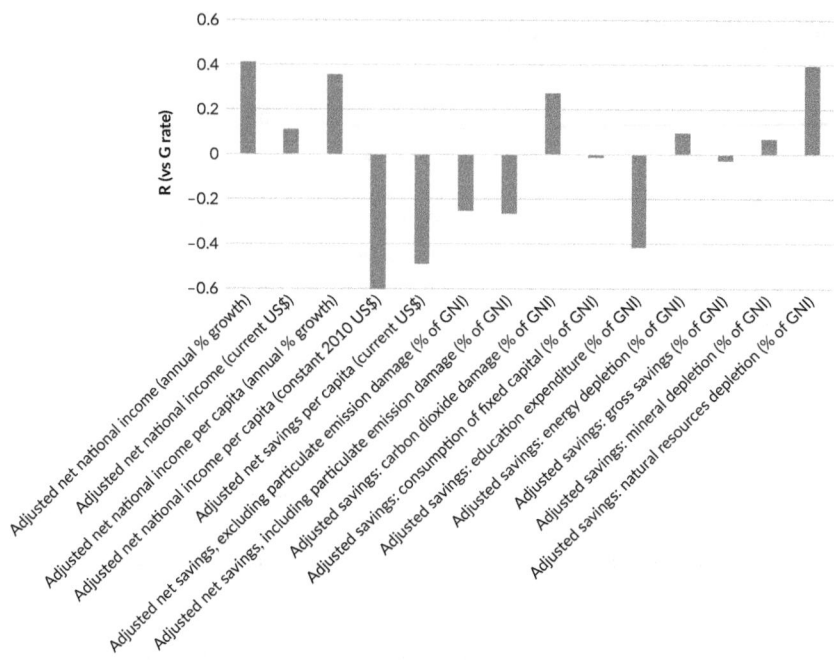

Figure 8. Bar chart showing the correlation (R, Pearson) between the incidence of the G allele of COMT in different populations and a series of financial indices related to economic growth and sustainability in the corresponding countries.

(Figure 8). No association was found with any of the adjusted net savings, excluding particulate emission damage (% of GNI), adjusted net savings, including particulate emission damage (% of GNI), adjusted savings: carbon dioxide damage (% of GNI), adjusted savings: consumption of fixed capital (% of GNI), adjusted savings: energy depletion (% of GNI), adjusted savings: gross savings (% of GNI) or adjusted savings: mineral depletion (% of GNI) (Figure 8).

Despite the high interconnection of the various financial indices used, such an approach unveils trends that may help us understand how these societies have been organized and operate, at least in economic trends. The most significant correlation was found between the presence of the G allele and the rate of sales tax in the corresponding countries. Sales tax is the tax imposed by the government or other administrative entities to support available infrastructure. The user of products that have passed a number of stages of manufacturing, pays it. Its level represents the collective outcome of various constituents consisting of both the actual societal needs, as determined by the specific way governmental support for goods and services has been structured in a given society and its overall acceptance by the corresponding population that is charged with this tax upon the usage of goods and services.

In countries where the corresponding populations had higher rates of the G allele, the sales tax rates were lower. Thus, the presence of the A allele for COMT (worrier allele) that has been linked to increased performance in complex environments at the population level correlates with the development of more complex societies which impose higher sales taxes to support infrastructure. The alternative is also possible and indeed, may happen simultaneously. In these countries, the populations are more likely to accept the incurrence of such taxes. Consistent with this trend is the significant negative correlation between income tax rate and the G allele prevalence. It suggests that higher taxation is aligned with populations exhibiting increased propensity for efficient operation in complex environments. The people in these societies also had increased income, saved more funds, and invested more in education. These populations also had a lower rate of depletion of natural resources, all of which, besides historical factors, may also be consequential to their ability to engage in long-term planning.

On the opposite hand, populations with higher prevalence of the G allele (warrior allele) have developed more collectivistic cultural traits characterized by increased altruism. Therefore, they may not depend highly on government-supported infrastructure. Similar to the protection offered by social support against depression, in these societies, assistance is provided in a manner that may not be centrally organized. Thus, what the government does not do systematically and formally, is done spontaneously by the individuals.

9.2. COMT and Governance

Beyond taxation and sustainability, correlating allelic frequencies for COMT is also informative for additional indices, reflecting how different societies are structured. The whole perception of Governance by the people reflects such an organization. For example, the Worldwide Governance Indicators reflect how people perceive Voice and Accountability, Political Stability and Absence of Violence/Terrorism, Government effectiveness, Regulatory Quality, Rule of Law and Control of Corruption in their countries. These indices are available for several different years and can be downloaded from www.govindicators.org.[25] It appears that the frequency of the COMT alleles correlates well with these indices.

For all of them, except Political Stability and Absence of Violence/ Terrorism, a tight negative correlation was seen with the frequency of the G allele of COMT (P<0.05) (Figure 9). In all cases, Japan and Korea were outliers exhibiting better governance indices than those predicted, while Pakistan and Bangladesh demonstrated the opposite trend, being outliers with worse than the predicted indices. In certain instances, other countries were outliers for different indices. European countries never displayed strong outlier behavior, while African countries and those from Central and South America only in rare instances were outliers.

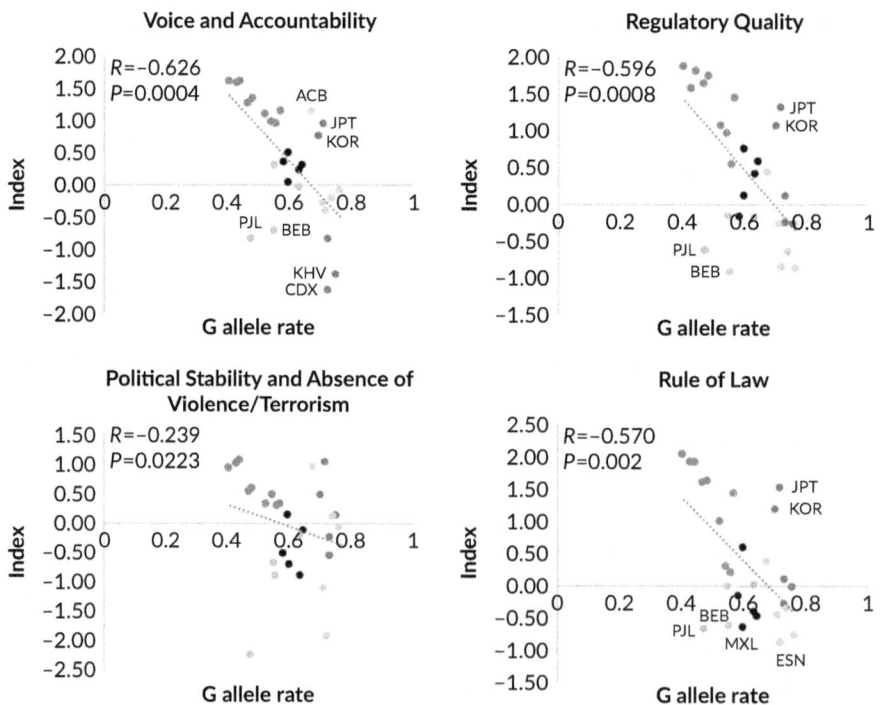

Figure 9. (Continued)

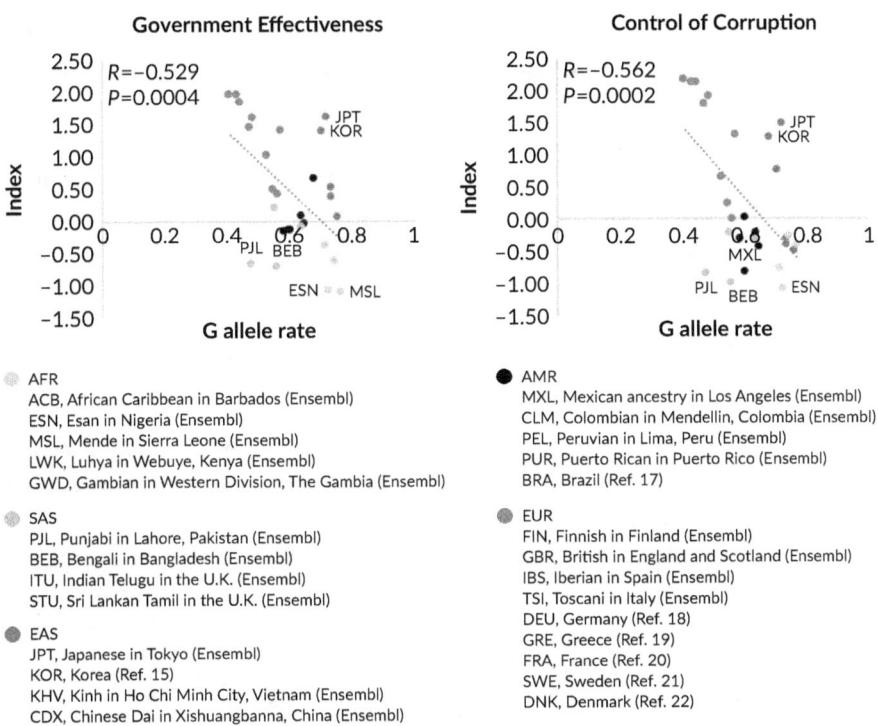

Figure 9. Scatter chart plot between the incidence of the G allele of COMT in different populations and a series of governance indicators in the corresponding countries. R and P values, (Pearson correlation) are indicated. The source of allelic frequency data is shown.

According to these Worldwide Governance Indicators the higher ratings for efficient governance were obtained for the European countries, for which a strikingly linear correlation with the frequency of the type of COMT allele was revealed. Considering that the models of governance applied in the corresponding indicators are primarily Westernized, the good fitness of the European countries in this model may also reflect this fact, that are indices developed by and for, the Western societies. On the other hand, the consistent behavior of Korea and Japan as positive outliers, performing better than what may be predicted, shows that the presumed impact of the genetic factors can be surpassed by appropriate policies. Furthermore, it demonstrates and the adaptability of different populations in different norms that have been introduced only relatively recently in their societies.

Obviously, the perception of good governance is rather subjective and reflects both the efficiency of governance and the perception of individuals. Furthermore, several genetic polymorphisms show geographical distribution similar to this observed for COMT. Thus, their high correlation with the governance indicators is expected. Yet, despite its limitations, this approach shows that biological and specific genetic factors may be linked to the perception of governance.

9.3. Oxytocin Receptor, Optimism and Self-Esteem

Oxytocin is a peptide hormone produced by the hypothalamus and released by the pituitary. It is the hormone that in women triggers the contractions at childbirth, and it is administered when physicians want to induce the process. It is also associated with several behavioral traits. For example, it elicits a feeling of calmness, and it is considered essential for the induction of bonding between the mother and the child as well as between couples.

Oxytocin produces its effects by interacting with specific receptors in the cell membrane, especially in the brain and other tissues. The oxytocin receptor is polymorphic, with several different alleles reported, many of which have been associated with pathologies. One specific variation designated as rs53576 is located in the third intron of the gene for the oxytocin receptor. According to this polymorphism, in some cases, individuals may have A while other times they may have G. The presence of G has been associated with increased sociality, self-esteem, and optimism (Li et al., 2015) and higher resilience to stress (Chen et al., 2011). By looking at Table 7, we realize that the frequency of the G allele in European ancestry populations is almost twice as high compared to the East Asians. Thus, G-bearing people, especially those in homozygosity, have a better view of themselves; they are more optimistic and social. Asians, on the other hand, in order to compensate for the lack of these personality traits, need to establish societies that can provide this support that on an individual basis is insufficient. Such civilizations are collectivistic communities in which what may be missing from the individuals, is provided by society. This appears to represent a similar response to the one that has emerged between the serotonin transporter and sociality. In both cases, collectivism provides to people what individuals feel is insufficiently perceived or received.

120

Table 7. Allelic frequencies for rs53576 (Oxytocin receptor) in different populations. Data were retrieved from the 1000 genomes project

	A frequency	G frequency
AFR	0.194	0.806
ACB (African Caribbean in Barbados)	0.182	0.818
ASW (African Ancestry in Southwest US)	0.270	0.730
ESN (Esan in Nigeria)	0.141	0.859
GWD (Gambian in Western Division)	0.190	0.810
LWK (Luhya in Webuye, Kenya)	0.227	0.773
MSL (Mende in Sierra Leone)	0.188	0.812
YRI (Yoruba in Ibadan, Nigeria)	0.185	0.815
AMR	0.356	0.644
CLM (Colombian in Mendellin, Colombia)	0.298	0.702
MXL (Mexican ancestry in Los Angeles)	0.422	0.578
PEL (Peruvian in Lima, Peru)	0.459	0.541
PUR (Puerto Rican in Puerto Rico)	0.284	0.716
EAS	0.650	0.350
CDX (Chinese Dai in Xishuangbanna, China)	0.608	0.392
CHB (Han Chinese in Beijing, China)	0.694	0.306
CHS (Southern Han Chinese, China)	0.657	0.343
JPT (Japanese in Tokyo)	0.654	0.346
KHV (Kinh in Ho Chi Minh City, Vietnam)	0.631	0.369
EUR	0.351	0.649
CEU (Utah residents, Western European ancestry)	0.298	0.702
FIN (Finnish in Finland)	0.465	0.535
GBR (British in England and Scotland)	0.379	0.621
IBS (Iberian in Spain)	0.308	0.692
TSI (Toscani in Italy)	0.313	0.687
SAS	0.449	0.551
BEB (Bengali in Bangladesh)	0.483	0.517
GIH (Gujarati Indian in Houston, TX)	0.369	0.631
ITU (Indian Telugu in the U.K.)	0.480	0.520
PJL (Punjabi in Lahore, Pakistan)	0.490	0.510
STU (Sri Lankan Tamil in the U.K.)	0.431	0.569

CHAPTER 10

Leaders and Followers

Societies need their leaders. During the earliest periods of human history, and concomitantly with the development of hierarchical societies, specific individuals were always acquiring leadership roles by guiding other people (the followers), obtaining a role surpassing that of their single existence and probably immediate family and taking responsibilities on matters involving a larger cohort of people or the group they belonged to. This role was reinforced by the political and social system by powers that at sometimes were based on metaphysical and at other times on more rational (usually in cases of democratic societies) concepts and arguments. Even before the emergence of these societies, during the tribal organization of *Homo sapiens*, certain individuals functioning as leaders were present, but with the corresponding administration being simpler, while both the criteria for establishing leadership and the responsibilities of the leaders were narrower. Regardless of the specifics, the emergence of leaders played a major role during human history, and that the qualities of such leaders, as well as a genetic tendency to that behavior, would be of particular interest.

Cholinergic receptor, nicotinic, beta 3 (CHRNB3) belongs to a family of receptors that mediate fast signal transmission at synapses. A polymorphism is located in the 5'-untranslated region (DNA sequence that is not translated into protein) of the corresponding gene designated as rs4950. The polymorphism refers to the presence of either G or A, and although it does not affect the primary sequence of the CHRNB3 gene, and thus of the corresponding receptor, it may affect its regulation and therefore its activity. This hypothesis is supported by the fact that adjacent to this position, several regulatory genetic elements have been identified. Earlier studies have provided evidence for a potential association between CHRNB3 polymorphisms and subjective responses and dependence to tobacco (Bierut et al., 2007; Saccone et al., 2007; Zeiger et al., 2008).

Besides this association of CHRNB3 and smoking habits, another intriguing observation is related to the geographical distribution of the rs4950 alleles. Table 8 shows the frequencies of G and A alleles in the polymorphism rs4950 in various populations around the world, according to the HapMap project.

While no difference in the frequencies exists between Asians and European ancestry populations, it is readily observed that all populations outside Africa, such as Asians, Europeans, Mexicans, and Indians, have high frequencies of the A allele. This frequency is greater than 0.7, which means that this allele corresponds to about 70% of all alleles in these people, as compared to below or around 0.3 or 30%, which is the frequency of the same allele in the African populations tested. The latter populations included representative samples from the Yoruban, Maasai, and Luhya,[26] as well as people with African ancestry living in the U.S.A.

Given the role of CHRNB3 in neural function, and thus its likelihood to affect behavior and decisions, this specific polymorphism is likely to regulate migratory behavior: Populations that resulted from migration events

Table 8. Frequencies of rs4950 polymorphism (G/A) in the CHRNB3 gene in different populations according to the HapMap project

Population	G allele frequency	A allele frequency
African ancestry in Southwest USA	0.728	0.272
Utah residents with Northern and Western European ancestry from the CEPH collection	0.226	0.774
Han Chinese in Beijing, China	0.277	0.723
Chinese in Metropolitan Denver, Colorado	0.202	0.798
Gujarati Indians in Houston, Texas	0.238	0.762
Japanese in Tokyo, Japan	0.128	0.872
Luhya in Webuye, Kenya	0.798	0.202
Mexican ancestry in Los Angeles, California	0.155	0.845
Maasai in Kinyawa, Kenya	0.689	0.311
Tuscan in Italy	0.173	0.827
Yoruban in Ibadan, Nigeria	0.844	0.156

(essentially all people outside Africa) have a high frequency of the A allele while African people, who apparently did not have to perform long-distance migrations from their original location in Africa, have a lower frequency in the A allele and a higher frequency of the G allele. Thus, the A allele may contribute to the acquisition of a migratory behavior by either inducing the decision to migrate or by offering an advantage to the people that have migrated.

This possibility could have been related to a rather intriguing hypothesis, namely attributing to the CHRNB3 gene polymorphism the function of the ultimate migration gene, surpassing a similar role discussed in earlier chapters for the DRD4 gene. Unfortunately, though, besides this superficial observation, no technically and statistically sound scientific studies—as yet, at least—are there to support this notion.

What has been reported however, is another interesting association between the rs4950 polymorphism and the emergence of leadership behavior. Specifically, individuals bearing the A allele were more likely to show leadership potential as compared to the G-allele-bearing individuals. Noteworthy, this report comes from the same investigators that earlier reported the interesting association between DRD4 and liberal political beliefs, namely Christakis and Fowler (De Neve et al., 2011), who found that the A allele of rs4950 is associated with leadership.

In this study, among the various scientifically acceptable ways to measure leadership, the investigators decided to use the occupancy of office associated with leadership as the actual measurement of leadership, which defines it in terms of formal and informal leadership roles that individuals maintain in their workplace.

Their approach was rather autobiographical and the ultimate question that they asked to the participants of their study was: "Thinking about your old job duties, which of the following statements best describes your supervisory responsibilities at your (current/most recent) primary job?" The potential answers included the following: "I (supervise/supervised) other employees" or "I (do/did) not supervise anyone."

Initially, by analyzing monozygotic and dizygotic twins, the investigators provided evidence that genetic factors play a role in the emergence of leadership, and then they proceeded to the genetic association studies by which they identified marker rs4950 as a candidate for leadership. Finally, they confirmed their results in a different, independent population, which adds particular validity to their arguments and conclusions.

Their findings are in line with the role of CHRNB3 in the regulation of the dopamine system, especially in view of the recognized role of this neurotransmitter in regulating behavior, although, as they recognized, a lot of additional work should be done in order to understand the neurophysiologic basis of these genetic association findings.

Taking the aforementioned association between rs4950 and leadership as a fact, the observation that the A allele is much more common outside than within Africa may also be viewed as a finding that is in line with the fact that the establishment of complex societal structures in the "out-of-Africa" populations, by definition, requires more (but probably not too many) individuals with leadership tendencies and capabilities. These potential leaders would have operated as such, at the original political, military, religious, economical, or any other aspects of the emerging societies. And the fact is that the more complex the societies are, the more "leadership" positions are developed and needed. By looking at the genotype data (the distribution of the A and G alleles in the individuals) for the same populations described in Table 6, we notice that an excess of AA individuals (both leadership alleles are present) exists (Table 9). Thus, more leaders than followers are present in these societies, an observation that contradicts the fundamental notion that in such hierarchical structures, leaders are less than the ones they will lead (the followers).

A likely explanation is that a leader in one domain of life is a follower in another. One can be—and most of the times actually is—a military leader and an economic follower at the same time, or concomitantly a political leader and a religious follower. Thus, the number of available leadership roles is high. This way, the attraction and tendency for leadership of the most possible people are being satisfied, the society operates smoothly, and turmoil is minimized. A fellow scientist, apparently with quite high self-esteem, once told me that if he was to be recruited for the priesthood, he would have been a pope by now! It's not uncommon to meet someone outside the context at which we recognize him as a leader and realize that he functions as an ordinary follower.

Various exceptions to the rule exist and certain individuals, especially during specific historical periods, have accumulated leading positions beyond a single domain of life. Thus, at the same time we see leaders in religion who also hold high political and frequently military offices as well. And not that uncommonly, they were extremely rich at the same time!

Table 9. Frequencies of rs4950 genotypes in different populations according to the HapMap project

Population	AA	AG	GG
African ancestry in Southwest USA	0.088	0.368	0.544
Utah residents with Northern and Western European ancestry from the CEPH collection	0.602	0.345	0.053
Han Chinese in Beijing, China	0.533	0.380	0.088
Chinese in Metropolitan Denver, Colorado	0.642	0.312	0.046
Gujarati Indians in Houston, Texas	0.564	0.396	0.040
Japanese in Tokyo, Japan	0.752	0.239	0.009
Luhya in Webuye, Kenya	0.046	0.312	0.642
Mexican ancestry in Los Angeles, California	0.707	0.276	0.017
Maasai in Kinyawa, Kenya	0.096	0.429	0.474
Tuscan in Italy	0.663	0.327	0.010
Yoruban in Ibadan, Nigeria	0.048	0.218	0.735

Such social and political structures, of course are currently viewed as oversimplistic and dangerous, and that they reflect less-advanced types of societal and apparently political organization. The current trend is towards diversification. Military leaders are distinct from the political or religious leaders, while crosstalk with the economical elite is considered at least suspicious.

Naturally, a positive feedback may operate between leadership and complexity: The existence per se and probable abundance of individuals with a tendency and desire to become leaders may not be just a simple result but might have also fueled the foundation of such complex societal structures that require many leaders to function properly. To that end, the leadership "demands" of an increasing number of individuals must be satisfied, and thus, the elevated complexity of the structures might have been necessary in order to generate various leadership niches sufficient to accommodate all prospective leaders. This may naturally represent a type of complex "social bureaucracy" but despite the potential dysfunctionalities it may cause, it has an undoubtful role in preserving social harmony.

Consistent with that notion, it is possible that building complex structures in all aspects of life is not a simple and sole requirement dictated from the advances of culture and civilization, but also an intuitive social strategy that unconsciously aims to satisfy the peoples' endogenous demand for leadership. More "leaders" found more complex societies that, in turn, require more leaders to function. Besides the social complexity, this interconnectivity also increases the cohesive bonds between individuals in a society, and thus, in part may compensate for the potential "deteriorating" consequences of individualism. As it becomes apparent, a positive regulatory network is being established between complexity and leadership at which the occurrence of one is not only a perquisite for the other but also induces it. We can better appreciate this by comparing the various civilizations and cultures that emerged worldwide throughout history with those in Africa that in general remained simpler, because probably fewer leadership roles had to be fulfilled.

Another interesting issue is the potential link between rs4950, leadership and migratory behavior per se. An interesting explanation for this link can be that migratory populations definitely demanded leaders. The success of these migratory waves was largely dependent on the abundance of people who could lead successfully both the expedition as well as the first communities that were established in the new territories. The more individuals who could be the leaders in such expeditions, the higher the chances for success for these newborn societies.

Furthermore, these leaders frequently were quite successful, reproductively speaking, since they represent good mating partners for cultural and other reasons. Thus, their genes were likely to be "over-represented" in the subsequent generations. In view of these observations, it is not that surprising that the leadership-associated allele of rs4950 is over-represented in all human populations outside Africa and under-represented in the African people.

CHAPTER 11

Eastern versus Western Traits Are "En Bloc"

We have seen only a few of the genetic traits that might affect human behavior. Furthermore, we were able to see that the frequencies of the corresponding polymorphic alleles differ between people of the major human cultures, namely those of the East and those of the West, or people of Chinese or European origin. Analogous differences, but at a smaller scale, also exist between other ethnic groups, but by comparing the two extremes, would likely be more informative in order to extract some interesting conclusions. Countless additional studies are also available in the scientific literature that either associate these particular loci to the regulation of various other behavioral traits or implicate loci other than those we have discussed in the regulation of human behavior. At this point, expanding the description and the analysis to additional traits and attempting to cover more genetic loci is probably useless, and it will only complicate things. However, some interesting points can be made.

Let's summarize here what we have seen until now:

First of all, among the various polymorphic genes that exhibit a relatively characteristic geographical distribution, namely differing in their frequency between East and West, at the same time also affecting certain behavioral traits, we have discussed DRD4, HTT, MAOA, COMT and Oxytocin receptor (Table 10).

Among the various different versions of these genes, the ones that are more common to Westerners have been associated with increased novelty-seeking, nomadic life, distance of out-of-Africa migrations, a tendency for financial risk-taking, and liberal ideology (DRD4). On the other hand, the various versions of MAOA and 5-HTT that are less frequent in the West predispose for collectivistic behavior, and as regards 5-HTT, this same version of the gene also increases the likelihood for someone to be happy with his or her life. Such collectivistic behavior at various instances seems to associate with the social support to individuals that protects them against stress.

Finally, the version of COMT that is seen frequently in the West predisposes for the "worrier's" as opposed to the "warrior's" behavior. In addition to those genes that have a "West versus East" geographical distribution, certain polymorphisms of the CHRNB3 gene are more common in all populations that are presently out of Africa, as compared to the populations in Africa, and also predispose for leaders', as opposed to followers', behavior.

Table 10. Examples of major behavioral traits that likely affect the development of Eastern and Western cultures and potentially have a genetic basis

Trait	Gene	Allele frequency
Novelty seeking	DRD4	7-repeat novelty seeking allele more common in the West
Migration	DRD4	7-repeat allele is associated with distance from Africa migration
Nomads/settlers	DRD4	7-repeat allele is associated with nomadic life
Political ideology	DRD4	7-repeat allele is more common in liberals
Financial risk taking	DRD4	7-repeat allele is more common in risk-takers
Individualism/ Collectivism	HTT	s allele (collectivistic) of 5-HTT is more common in the East
Happiness	HTT	l allele has higher prevalence in individuals happy with their life
Individualism/ Collectivism	MAOA	3-repeat allele (collectivistic) more common in the East
Warrior/Worrier	COMT	A-allele (worrier) more common in the West
Altruism	COMT	G-allele (warrior) associated with altruism
Leader/Follower	CHRNB3	A-allele (leader) more common in populations Out-of-Africa
Optimism, self esteem	Oxytocin receptor	A-allele (low esteem and optimism) more common in Asians

An interesting observation from these results is that it appears that these independent polymorphisms appear to "fit" together quite well in shaping someone's personality. In other words, we can identify a sort of "complementation" and synergy between the various traits contributed by these polymorphisms seen in the same population. For example, it is more likely for a novelty-seeker to exhibit individualistic behavior, rather than collectivistic. Being a novelty-seeker ultimately means committing to a constant process that aims to satisfy one's individual needs, an attitude that is intrinsically related to individualism. The same applies for the political ideology at which being a liberal, "fits" better with being individualistic rather than collectivistic in his or her behavior in the perception of life. Liberal as opposed to a conservative ideology may imply that someone wants to alter a certain status quo that in turn is usually intrinsically linked to an established balance and harmony in a society. In that case, the individual needs (the needs of the individuals) have been displaced by those that benefit the society as an entity. Not of course, that this balance is already the best possible, but it is still capable of offering the security of belonging to a specific group. Furthermore, the process of this change and alteration, as such, possesses a component of risk-taking against a speculative novel status that is unknown as yet, and thus, poorly defined. In addition to those tendencies, being a "worrier" as opposed to a "warrior" means that someone has the ability to adapt at the complex conditions, which are usually the conditions faced by the novelty-seekers in the new environments they have to face and deal with. This is also analogous to the liberal ideology at which the complex process of change, as such, in general, requires the ability for adaptation in such novel conditions, which in turn is benefited by the "worrier's" behavior. The association between having a tendency for altruism and collectivistic behavior is also self-evident, since the benefits of altruism become apparent only in societies with an increased collectivistic component. On the other hand, a worrier that is characterized by reduced impulsivity, as compared to the warrior, is likely less altruistic, since strictly speaking, the latter attitude does not satisfy directly his or her individualistic needs. As we have discussed, an individualist behaves altruistically only when the benefits of such behavior are visible. In that sense, he is an altruist by a manner that is individualistic(!). Finally, it appears that genes that at the level of individuals predispose them to stress are more common in the populations with collectivistic societies. Such societies have established norms that can provide social support to individuals that are more vulnerable and protect them.

An interesting exception, at least if we view things quite superficially, to those observations is related to the association of the happiness allele of 5-HTT with the polymorphism that is more frequently seen in Westerners. The "change"—associated trends of novelty-seeking, liberal behavior, or risk-taking, and even that of the "worrier's" behavior, probably contain, in a latent status, the component of dissatisfaction. If you want to change things, it means that you are not satisfied with what you have already. That, in turn, may correspond to and nourish unhappiness. Consistent with that view, it is hard to understand how one is happy with his life and at the same time seeks to change it. Moreover, his willingness to take risks for that change increases his potential dissatisfaction even more. It is possible though, that either the same exact component of change is the one that contributes to the perception of happiness, or some other benefits of this "happiness"–associated polymorphism, such as the stimulation of individualism that is also associated with 5-HTT—are the ones that prevail. In that case, of course, we reach the surprising conclusion that the "default" and spontaneous condition is not that of happiness, but the latter emerges as a side-consequence of other benefits conferred by the corresponding polymorphisms. You have to be happy because, in principle, you are an individualist!

Despite those peculiarities, an unbiased conclusion that can be reached is that the sum of the aforementioned polymorphisms and the corresponding behavioral trends they control appear to fit together quite well. In other words, they appear to behave "en bloc," which implies a certain degree of complementarities in shaping up the major Western and Eastern cultures and the archetype personalities of the corresponding people.

Such a notion is different from the genetically defined and measurable linkage, according to which two genes, or genetic loci, by being closely located in the same region of the chromosome, are inherited together. Thus, we may see that the physiologically irrelevant trait A and trait B are frequently seen together, just because the corresponding genes controlling them are closely located in the same chromosomal region.

In the present case, though, the genes discussed earlier, which are controlling these behavioral traits, are located in different chromosomes and/or distal chromosomal locations, and thus, no genetic linkage exists. Of course, genetic founders'—related phenomena that are associated with the original shaping of the genetic structures of the corresponding populations—have naturally occurred. In that sense, what we witness today is just the reflection of the original genetic composition of the people that inhabited the corresponding areas at the first instance. But even in view of that fact, that the corresponding behavioral trends are related is interesting.

A plausible explanation is that pure randomness (or coincidence), supplemented with a specific dose of relativity, contributed to this observation. What we see as reasonable and logical now is the collective outcome of a given, continuous, and distinct process that has already shaped our minds accordingly in order to make us view things in a certain way: The reasonable way! Our civilization has pre-defined our way of thinking to interpret certain and specific attitudes (and actually whole sets of attitudes) as reasonable and sensible. To that end, it is quite hard for us to imagine that an alternative—and not simply the "opposite," consistently with Eastern-Western dualism—cultural line and behavioral pattern could exist and be functional. It is the same difficulty we have as lay people and not theoretical physicists in understanding worlds and phenomena that operate in five or six dimensions. And, indeed, the more isolated intellectually we are, the higher the difficulty to perceive such alternative possibilities is.

In theory, if we were able to have large populations at which these combined traits, such as both Westerners' novelty-seeking allele of DRD4 and Easterners' collectivistic allele of HTT, represented the majority, it might have been quite informative. It would have allowed us to observe the features of the corresponding cultures and whether they would have become an interesting hybrid between what we currently perceive as East and West or a novel, third- and fourth-type culture, distinct from the aforementioned two (Figure 10). In that latter case, of course, the whole concept of the dualistic perception deeply carved into our minds would have suffered from severe damage.

Consistent with such a hypothetical scenario, it would be even more interesting to look at what we understand as "en bloc." Would the corresponding traits, in such a case, still feel like they fit well together?

Certain subpopulations, though, and groups of people within these major populations, do exist. The frequencies of the polymorphic sites are not absolute—neither 100% or 0—for any of the traits we discussed earlier. Actual frequencies are between these values and, thus, all possible combinations occur.

Thus, the Westerners' novelty-seeking allele of DRD4 and the Easterners' collectivistic allele of HTT do co-exist in many individuals. The only

133

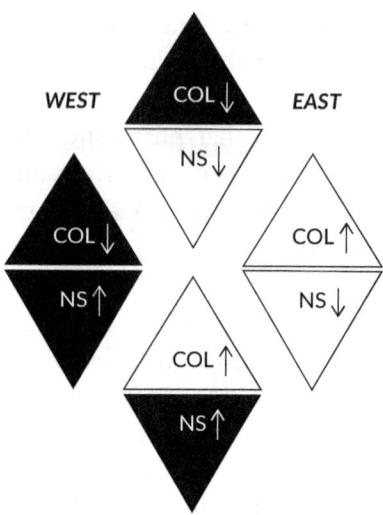

Figure 10. This diagram depicts the *"en block"* traits collectivism (COL) and novelty-seeking (NS) in the East and the West. The upper and lower panels depict mixed populations at which high collectivism is accompanied by high novelty seeking and *vice versa*, low collectivism is accompanied by low novelty seeking behavior. It is noted that neither collectivistic nor novelty seeking societies are homogenous, as depicted here. Furthermore, the frequency of collectivists in individualistic societies and novelty averters in the societies at which novelty seekers prevail are not equivalent.

difference is that, since in the West the novelty-seeking and individualistic alleles are at a higher frequency, the chances to co-exist are higher; thus, the number of these individuals is greater. However, the fact is that such people, with combined Eastern and Western traits are present in both societies and are actually integral parts of these cultures that are dominated—in quantitative aspects—by the "en bloc" traits.

That the individuals with "combined" traits are just the minority likely masks their inherent and potentially distinct features. In other words, since it is the majority that makes the rules and defines the social norms, these low-novelty-seeker individualists, or high-novelty-seeker collectivists, by being a minority, miss their chance to make their point and leave their cultural fingerprint by establishing a distinct cultural trend-line.

Another possibility is that the "en bloc" characteristics we observed indeed helped the diversification of these cultures. There is a certain theoretical possibility that the result is not due to quantitative but rather qualitative processes. To that end, cultures developed by such "hybrid" populations, with "mixed" (the Easterners' and Westerners') predominant traits (for example, by being concomitantly novelty seekers and collectivists), might have been unstable, and eventually they might have been absorbed by the cultures that exhibited the corresponding characteristics "en bloc." We may understand that by considering a society that is dominated by such novelty-seekers and collectivists. These individuals that represent the majority in these populations will constantly have to compromise the risks associated with the novelty-seeking behavior with the stability of the collectivism. The continuous compromise of these opposite tendencies is possible that might have nurtured a conflict. It is quite likely that, in the long run, such conflict would not have survived. Imagine, for example, that at a period of a certain massive disaster that required the advantages of collectivism, the society would, at the same time, also have to cope with the needs of the majority of individuals, being the risk-takers and novelty-seekers (that also display a tendency for liberalism). This state would probably have been contradictory, suffering from some kind of intrinsic instability.

Imagine another hypothetical society at which the majority is composed of individuals that at the same time are individualistic but not novelty-seekers. As individualists, they'll have to pursue happiness in view of the needs of the single person. But without being capable of taking certain risks and being novelty-seekers, their capacities will be limited, and such societies would not have advanced at a degree comparable to that of the societies with these traits "en bloc." Individualism would not be able to function smoothly and efficiently in the absence of a tendency for novelty-seeking. The intrinsic need for being different from the others, which is an integral component of the individualistic societies, in order to be satisfied, requires the tendency for novelty-seeking and risk-taking, as well as a certain degree of liberalism within the context of the doubt of the existing "status quo." In financial terms, this is analogous to the widely accepted notion that economic stability depends on growth.

Even if such societies existed (Figure 10), we may anticipate that in the long run they would not have been as successful as the ones at which the aforementioned characteristics existed "en bloc." Probably, the subpopulations with the corresponding characteristics presented "en bloc" may have eventually prevailed, resulting in societies that simulate the dualistic present-day Eastern/Western discrimination.

Of course, another important point that needs to be made is that despite that such individuals with "combined" traits represent the minority in both the East and the West, they still contribute significantly to the intellectual and behavioral variability of the corresponding populations and cultures, a condition that is imperative for their advancement. Another intriguing possibility is that, indeed, populations and associated cultures with such mixed characters existed in the past, but they might have been eventually absorbed by these major Eastern and Western cultures that dominated human history. To that end, by definition, they were classified in relation, and actually in between, these two extremes.

11.1. Some Personality Traits Are Linked, but Not Genetically, in Easterners and Westerners

Since many polymorphic alleles linked to personality traits co-exist among the Easterners and the Westerners, we could apply a reverse approach to explore which of these variants tend to show similar frequency in different populations worldwide. This would probably provide some insight on whether the co-existence of some of them in specific populations is random or if it reflects something deeper. Furthermore, it could show if there is some genetic basis explaining the co-existence of different behavioral traits in the different cultures.

Table 11 shows several polymorphisms in single base pairs (SNPs) that have been previously associated with various personality traits. These polymorphisms could differ in their frequencies between the Easterners and the Westerners, but even if they don't, their frequency is variable between different populations worldwide. By using these frequencies from 26 societies worldwide, data were retrieved [from Ensembl (1000 Genomes Project, phase 3)], and the correlation coefficient was calculated for all possible pairwise comparisons (Figure 11). This way, how tightly the frequency of one polymorphism for different populations correlated with the frequency of another, in the same populations, could be assessed.

Some polymorphisms had a high correlation with each other (in their frequencies), while others did not. However, since none of these polymorphisms were genetic, whether they co-existed was irrelevant from their genetic co-segregation. The question now is whether the ones that exhibit variable but correlated frequency worldwide induce personality traits that are meaningful when they are present together. Does this imply that they

Table 11. Genetic locus and chromosomal location of SNPs that have been associated with personality traits in various analyses and metaanalyses

SNP	Locus	Chromosomal location	Personality trait	Reference
rs11584700 (A/G)	LRRN2	1:204607855	Education attainment	Rietveld et al., 2013
rs4851266 (C/T)	LOC150577	2:100202017	Education attainment	Rietveld et al., 2013
rs2024488 (A/G)	LOC101928250	2:216798245	Extraversion	den Berg et al., 2016
rs7600563 (T/G)	CTNNA2	2:80488024	Excitement-seeking	Terracciano et al, 2011, Sanchez-Roige et al., 2017
rs57590327 (G/T)	GBE1	3:81952561	Extraversion	Lo et al., 2017
rs1865251 (A/C)	CADM2	3:85439919	Risk taking	Boutwell et al., 2017 Day et al., 2016
rs53576 (G/A)	OXTR	3:8762685	Sociality	Kim et al, 2010
rs1477268 (T/C)	RASA1	5:87132049	Openness	De Moor 2010
rs3814424 (C/T)	LINC00461	5:88673135	Conscientiousness	Lo et al., 2017
rs214618 (C/T)	PTPRD	6:18197988	Openness	Kim et al., 2013
rs1049353 (C/T)	CNR1	6:88143916	Stress	Hill et al., 2013
rs9320913 (C/A)	LOC1001129158	6:98136857	Education attainment	Rietveld et al., 2013
rs2164273 (A/G)	MTMR9	8:11310990	Extraversion	Lo et al., 2017
rs4950 (A/G)	CHRNB3	8:42697490	Leadership	De neve et al., 2013
rs9650241 (T/G)	chr. 8	8:72115189	Agreeableness	Bae et al., 2013

(Continued)

Table 11. (*Continued*)

SNP	Locus	Chromosomal location	Personality trait	Reference
rs6481128 (A/G)	PCDH15	10:54901602	Extraversion	Lo et al., 2017
rs1800955 (T/C)	DRD4	11:636784	Novelty seeking	Okuyama et al., 2010, Ronai et al., 2001
rs1800497 (G/A)	DRD2	11:113400106	Memory and attention	Söderqvist et al., 2014
rs1426371 (G/A)	WSCD2	12:108236003	Extraversion	Lo et al., 2017
rs8010306 (A/G)	SLC25A21	14:36680955	Extraversion	Sanchez-Roige et al., 2017
rs7498702 (T/C)	RBFOX1	16:6477347	Extraversion	Lo et al., 2017
rs2576037 (C/T)	KATNAL2	18:47059049	Conscientiousness	De Moor 2010
rs117292860 (C/A)	DOT1L	19:2227622	Extraversion	Sanchez-Roige et al., 2017
rs4680 (G/A)	COMT	22:19963748	Worrier/warrior	Montag et al., 2012, Hashimoto et al., 2017

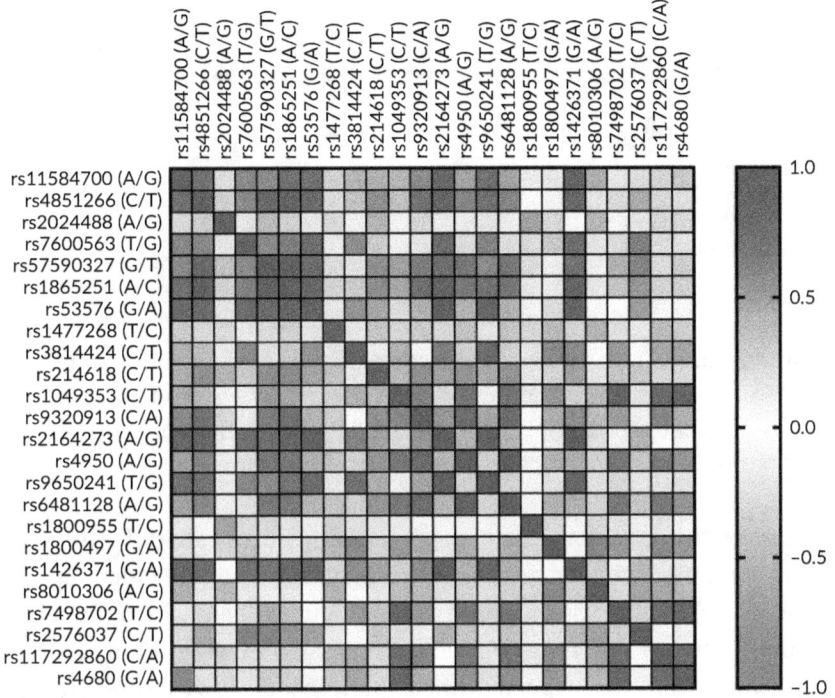

Figure 11. Heat map showing the correlation, for all pairwise comparisons, between SNPs associated with behavioral traits. Allelic frequencies were obtained from 26 populations worldwide by using data obtained from Ensembl (1000 Genomes Project, phase 3).

may operate in a concerted manner contributing to personality traits that exhibit complementarity, or at least non-exclusiveness?

Since East Asians and Europeans are more diverged culturally, comparing those would make sense if we wanted to unveil such variations that contribute to cultural differences. Among all these polymorphisms some differed considerably, while others did not, in their frequencies between East Asians (EAS) and Europeans (EUR). Five SNPs were selected, that had the higher difference in allelic frequencies between EAS and EUR that ranged from 0.273 to 0.358. The specific SNPs were rs7600563, rs53576, rs3814424, rs2164273, and rs1426371 and the corresponding genetic loci were CTNNA2, OXTR, LINC00461, MTMR9, and WSCD2. When the correlation coefficient was calculated only for those SNPs, for all 26 experimental populations worldwide, a highly significant correlation (P<0.05) was detected for all pairs (Figure 12a).

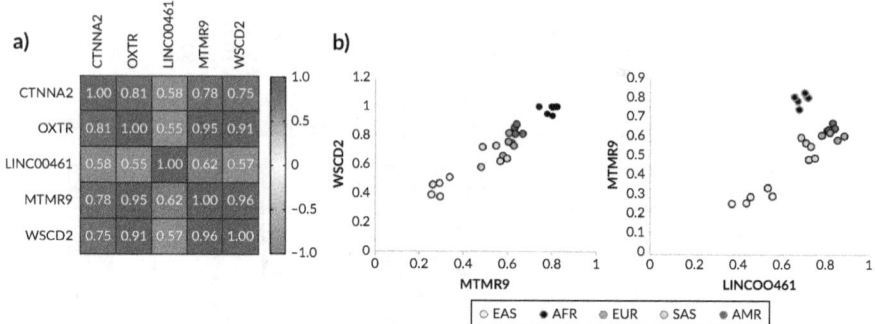

Figure 12. a. Heat map showing the correlation in allelic frequency, between the EAS and EUR populations, for the selected five SNPs identified in (a). **b.** Pairwise scatterplots between the allelic frequencies for representative SNPs (genetic locus is indicated) in various populations.

Hence, the genes for which East Asians and Europeans differ more have highly correlated frequencies in several populations worldwide. Representative pairwise comparisons are shown in Figure 12b, in which the relevant distance between different populations can also be seen. One can notice divergences from the expected linear patterns, such as the polymorphic pair on the right panel of figure 12b. In this case, the African populations were outliers suggesting that not in all cases a linear profile is recorded, which in turn may also tell us that when it does it is likely meaningful.

Thus, a set of polymorphisms, especially those that differed more between EAS and EUR, were increasingly correlated in their frequencies in different worldwide populations. This does not seem to reflect the behavior of all polymorphisms because others did not correlate well.

There are many mathematical algorithms available to cluster together groups of specimens that have several quantitative variables recorded. These tools are used for different types of biological analyses, for example, when we want to compare expression profiles between samples and see which ones are more similar. If we apply this approach to variation data by using the different polymorphisms and their frequencies in the populations as variables, we could cluster them together according to their similarities. We could identify this way which ones exhibit more similar patterns of change. If, for example, one polymorphism has a frequency of 10% in one population, 50% in another, and 30% in a third, it will cluster closely to a polymorphism that has 20%, 90%, and 55%, respectively, at least compared to a polymorphism that has frequencies of 10%, 15%, and 95%.

In parallel, we could group not only the polymorphisms but also the populations to see which ones are more or less similar with each other. By doing this, we can come up with "dendrograms" or "trees" that depict "distances," similarities, and differences between polymorphisms and populations.[27]

An example of such tree is shown in Figure 13. On the vertical axis, the allelic frequencies accurately clustered together with the different population groups according to their genetic similarity, with the one in AFR exhibiting the highest divergence from the other groups. EAS and SAS (South Asians) populations grouped together, as well as AMR (American populations) and EUR, which is in fairly good agreement with the patterns of historical migration and the chronological divergence of these groups.

The only discrepancy was with the PUR (Puerto Rican) population that clustered closely with EUR and not with the AMR populations, as would have been expected from the historical migration information we have. This most likely reflects a limitation related to small number of polymorphic loci used in this analysis, possibly resulting in misleading conclusions. Nevertheless, it appears that it is informative enough as it can cluster all but one of the different populations with high accuracy. It has to be mentioned though that the specific history of the Caribbean groups is still inconclusive. Genomic data appear controversial since, according to some studies Caribbean populations have been established, almost exclusively, by populations that migrated from Northeastern South America and Central America (Fernadez et al., 2020), while according to other studies, North American ancestry have contributed to their genetic makeup (Nägele et al., 2020).

As regards to the specific SNPs, they were classified into two major branches: The first involved SNPs in the vicinity of SLC25A21, DRD2, DRD4, RASA1, LOC101928250, and PTPRD, while the second involved all other SNPs. In turn, this second branch was further divided in two more branches; the first of which involved the SNPs that are located in the vicinity of CTNNA2, OXTR, LINC00461, MTMR9, and WSCD2 that showed the highest difference in their frequency between the EAS and the EUR populations.

Each specific experimental strategy may provide different gene sets with overlapping or not gene targets. And depending on how detailed, accurate, and informative the parameters used are, the power of such analysis will be higher. Regardless of their specific identity, the strategy used

141

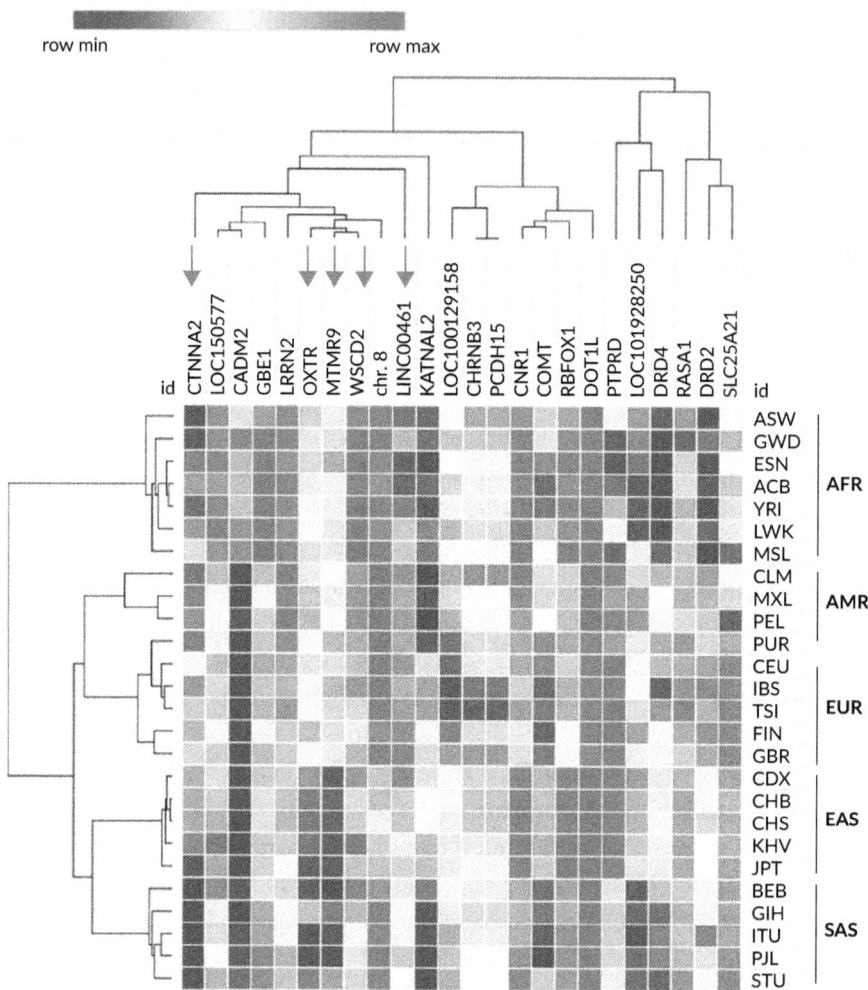

Figure 13. Unsupervised hierarchical clustering based in the allelic frequencies for all SNPs (genetic locus is shown) associated with personality traits (horizontal axis). In the vertical axis the corresponding populations are shown. The arrows indicate the loci for which highest difference in frequency between EAS and EUR populations exists.

here showed that subgroups of polymorphisms that are associated with personality traits have the tendency to show similar frequencies in different populations, high intermediate or low, while others do not.

Does this clustering mean something for the behaviors linked to them? It probably does because the personality traits that are associated with the

SNPs that coexist in the different populations show complementarity in establishing trends that are aligned with the cultural characteristics of the corresponding societies (Table 12). For example, in EAS populations, the allele contributing to reduced excitement seeking and increased conscientiousness (the tendency for responsibility, organization, hard-working, goal-directed activities) are all features that are recognized as being more typical in the collectivistic Eastern societies. Alternatively, in the EUR populations, the alleles that predispose for high excitement seeking and reduced conscientiousness predominate, a trend that appears to be compatible with the organization of the Western, individualistic societies. Especially regarding conscientiousness, this appears to be an essential factor for the collectivistic societies that rely on increased interdependence. In addition to those, the allele for rs53576 in the oxytocin receptor (OXTR), which has been associated with lower empathy and higher perception of loneliness, predominates in the EAS populations, synergizing with the development of a more collectivistic society at which collectivism provides compensatory and supportive environment against social stress superseding what is provided (and received) by the people on an individual basis.

As regards to the alleles for rs2164273 and rs1426371 that are associated with extraversion, a significant difference also exists in the frequencies of Western and Eastern populations. In this case though, it appears that the G and A alleles respectively, that appear to influence positively extraversion, predominate in the Easterners. This appears to contradict the reduced extraversion scoring exhibited by the Easterners as compared to the Westerners (Schmitt et al., 2007). A plausible hypothesis is that enrichment of certain alleles has occurred to provide compensatory protective mechanisms in these populations for a trait that has been linked to several additional polymorphisms, beyond the specific two (Table 12).

As to their implications in establishing personality traits, these polymorphisms were also shown to predispose for tendencies that are aligned with our accepted notion for the historically attained cultural divergence between these populations. For example, variations that independently contribute to reduced excitement seeking increased conscientiousness and the perception of loneliness. They were increasingly correlated with each other in different populations and predominate in East Asians. This likely contributed to the enhanced collectivism of this society. In the Westerners, whose cultural traits developed in the opposite direction, the corresponding SNPs showed lower frequencies.

Table 12. SNPs and personality traits in Eastern (EAS) and European ancestry (EUR) populations

SNP (locus)	Personality Trait	Population		Reference
rs7600563 (CTNNA2)	Excitement seeking	East (G) Low excitement seeking	West (T) High excitement seeking	Terracciano et al., 2011
rs53576 (OXTR)	Empathy, loneliness	East (A) Low empathy, High loneliness	West (G) High empathy Low loneliness	Kim et al., 2010
rs3814424 (LINC00461)	Conscientiousness	East (T) High Conscientiousness	West (C) Low Conscientiousness	Lo et al., 2017
rs2164273 (MTMR9)	Extraversion	East (G) High extraversion	West (A) Low extraversion	Lo et al., 2017
rs1426371 (WSCD2)	Extraversion	East (A) High extraversion	West (G) Low extraversion	Lo et al., 2017

It is plausible that genetic founder effects early during the historical migrations and establishment of these populations at distal geographical locations caused these variations in SNPs' frequencies. These distinct compositions of alleles had established the tendencies for specific personality traits in these populations and ultimately have shaped the corresponding cultures as they evolved during the historical period. In various instances, exemplified by the prediction for higher extraversion in Easterners, such tendencies may operate oppositely from what we see now, which, in turn, suggests the establishment of a compensatory mechanism that can "fine-tune" specific outcomes. Nonetheless, it appears likely that clusters of variations co-exist in the different populations, functionally operate in a manner that is consistent with their non-genetic linkage, and which synergize to shape cultural processes that are expressed at the level of the society as a whole.

Such effects appear essential in facilitating the establishment, maintenance, and re-enforcement of the cultural identity of the different populations. By differentiating them sufficiently at the genetic level, the differences in allelic frequencies may cause analogous differences in the quality and the intensity of personality traits that ultimately cause the establishment of "coherent" cultures in the different ancestry populations. The East-West divergence is the most prominent example of this, in which genetic differences, due to an early divergence of societies, are maximal, and therefore the associated with them personality traits are more visible and easier to track. What we witness today, exemplified in the cultures of the East Asians and the Westerners, is the successful outcome of multiple attempts of various populations to move out of Africa and subsequently establish successful cultures around the world. It is plausible that other additional attempts have been made but were not successful, and this can be partly due to the genetic make-up of these groups that were unable to establish a viable cultural fingerprint that could persist in history. With that, I don't mean only the failed attempts of people to establish successful settlements at distal locations and probably shortly after their establishment seized to exist because of lack of food, hostile environment, and other adverse contextual factors. I also mean the cases in which biological settlements could have been successful but were not because they were absorbed by other populations without leaving a significant cultural fingerprint.

The presumed significance of being able to establish a coherent culture may also contribute to explaining why in Africa, despite the astonishing genetic diversity, no major civilizations have emerged throughout history. By not being the subject of significant genetic bottlenecks—due to the lack of migration and the related to them founder effects in the migrant groups– the establishment of coherent behavioral traits and cultures were not favored. However, when people are present in populations in which they do not reflect the majority of the genetic pool, as individuals, they may thrive, as they frequently do so!

11.2. Divide and Conquer: The Genetics of Social Networks and their Impact in Social Organization in East and West

The coach of the soccer team in my hometown had recognized that the team's bad performance in a recent tournament was due to the lack of coherence between teammates. They were trying to follow his instructions but without real motivation and innovation in their playing tactics. He started lecturing players about it, but not much was accomplished. He decided to implement more drastic measures to bring them together. One day he came up with a different approach: He required each player to take a cold shower before and after each rehearsal. Everybody started complaining and accusing of inhumanity, madness, and all that. Over time this progressed, and some of them even complained to the team's administration. They established a small "opposition" subgroup, collected signatures from other team members, and requested the replacement of the coach. They even started now doubting his advices on the strategies he proposed during the games, confronting him and criticizing his instructions during the team's practices. Each player had the backing of the other players now. All these changes eventually affected their relationships with each other. They even started meeting after the practice to go to the movies and hang out together. Something that was inconceivable before the cold shower policy was implemented.

Team members identified in the coach a common "enemy." With this mandatory, repeated cold shower task, they came closer to each other and started functioning as a more coherent group. New friendships were established, the social dynamics of the team changed, and even the authority of their coach as the team's leader was doubted. The coach took advantage of the "divide and conquer" paradigm, and he applied it, but from the opposite direction, with the risk of compromising his authority. He united them by rendering himself as their "enemy." Social connections and the strength of friendships in the group had an impact on how authority was viewed. The coach was not their friend anymore, but they became closer friends with each other, started doubting their coach, criticizing him, and even wanting to substitute him.

A few years ago, Christakis and Fowler, by studying friendship in social networks, showed that genetics can contribute to our selection of who our friends will be (Fowler et al., 2011). Specifically, they discovered that the SNP rs1801272 in gene CYP2A6 was associated with heterophily. That means that no connections between individuals having the minor allele of CYP2A6 were recorded. The opposite trend was found for DRD2 that exhibited homophily (the investigators recorded the presence of significant clusters of people having similar genotypes for DRD2). Since our genes may influence who our friends can be, this may also impact how society is structured. After all, society is the network of connections by which a group of people is organized. More specifically, it emerges as a possibility that the percentage of each allele in the population, and the abundance of the different genotypes, may play some role in the long run in determining how abundant, tight, and complex the clusters of social connections will be in different societies at which the populations have different frequencies for these genes.

Let's now see what the frequencies of these alleles are in the EAS and the EUR populations. According to ENSEMBL, rs1801272 for CYP2A6 has two alleles, with A being the major (common) and T being the minor (rare) allele. These are having a frequency in EUR of 97% and 3%, respectively. The T allele is completely absent from the EAS where individuals have exclusively allele A. With regards to the frequencies of the genotypes it seems that 93% of the EUR are AA and 7% were AT, while no TT individuals were detected in this ENSEBL's population set.

As regards now the polymorphisms for DRD2, the frequencies of rs1125394 in EAS populations are about 58% for T and 42% for C, while the corresponding frequencies in EUR are 85% and 15% respectively. The corresponding genotypic frequencies in EAS are 34% (TT), 18% (CC) and 48% (TC) while in EUR it is 74% (TT), 3% (CC), and 23% (TC) respectively.

To model how these allelic frequencies influence the bonds in the society we calculated the number of social connections in two hypothetical populations of 1000 individuals that had genotypic frequencies similar to those predicted by Ensembl for the EAS and the EUR populations. For this modeling we assumed the following:

1. Each individual in the population will be able to establish a connection with each and every other individual with equal probability.
2. Homophily (DRD2 polymorphisms) has 100% penetrance. Therefore, no connections develop between individuals with different genotypes but only between those of the same genotype.
3. Heterophily (CYP2A6 polymorphisms) has also 100% penetrance. That means that no individuals with the minor CYP2A6 allele establish connections with individuals bearing the same, minor allele.

By applying these criteria, we calculated the number of social connections by the equation

$$C = (n-m)(n-1) + m(n-m)$$

In this equation,
C reflects the number of social connections
n is the total number of individuals in the group (population size)
m is the number of individuals with the social network disruptor genotype (AT here which is present in EUR only).

In this equation, for simplicity, each directional association was considered as an independent connection. For example, if two individuals are connected, this connection will count as equals to 2 (one originating from each individual).

By applying the equation for a population of 1,000 people, when no genetic impact was considered (all connections form independently of the genotypes), 999,000 social connections were predicted among all individuals. When only the contribution of the CYP2A6 was considered, then for the EUR population the number of connections decreased to 994,170, while for the EAS in which the minor allele is absent the number of connections remained the same (999,000).

When however, homophily was taken into consideration then in the 3 genotype groups the number of connections in the EAS was 115,260 (TT group), 32,220 (CC group) and 229,920 (TC group) (equals to 377,400 total connections). In the EUR population the corresponding number of connections were 546,860 (TT), 870 (CC) and 52,670 (CT) (600,400 total connections) (Figure 13). When both homophily and heterophily were taken into consideration then the connection numbers became 115,260 (TT) 32,220 (CC) and 229,920 (CT) (377,400 total connections) in the EAS. In the EUR population these numbers were 544,228 (TT), 867 (CC) and 52,426 (CT) (597,523 total connections) (Figure 14).

It seems that especially homophily profoundly affects the number of social connections in a population. To explain the consequences of these predictions, we can hypothesize that the abundance of social connections between individuals may inversely correlate with the rigidness and the collectivistic organization in the societies, especially in terms of how authority is viewed. It is plausible to speculate that increased connectedness between individuals in a society will compromise the collectivistic culture, particularly in relation to their higher administrative authority. In that case, society tends to become more complex, and as a consequence of this, the social stratification is now less rigid as social cohesiveness likely becomes weaker. This is a consequence of the fact that the complex connections between individuals in EUR likely supersede the power of societal bonds, promoting individualism.

On the other hand, the genetic makeup of the populations in East Asia, where the frequencies of the DRD2 allele promote fewer social connections among the individuals due to homophily, contributes to the emergence of a more collectivistic society in which the authority is doubted less severely. By having fewer connections between them, individuals in this community can be more adherent to the norms instructed by society as a whole.

These tendencies, which are presumably guided by variations in DRD2, are reinforced by the polymorphisms in the CYP2A6 gene which are present in the EUR population only. Albeit at low frequency and acting by modestly reducing the number of social connections in the EUR population by triggering heterophily, these variations may have another effect: They may contribute to the establishment of clusters operating as "social disruptors" since they establish some forms of social barriers between individuals and presumably clusters.

Like in the example with the soccer coach, strong social connections may compromise the coherence of the societal structure. Especially the relationship between the group members and the authority, which now may be doubted and questioned. We can view the collectivistic society

149

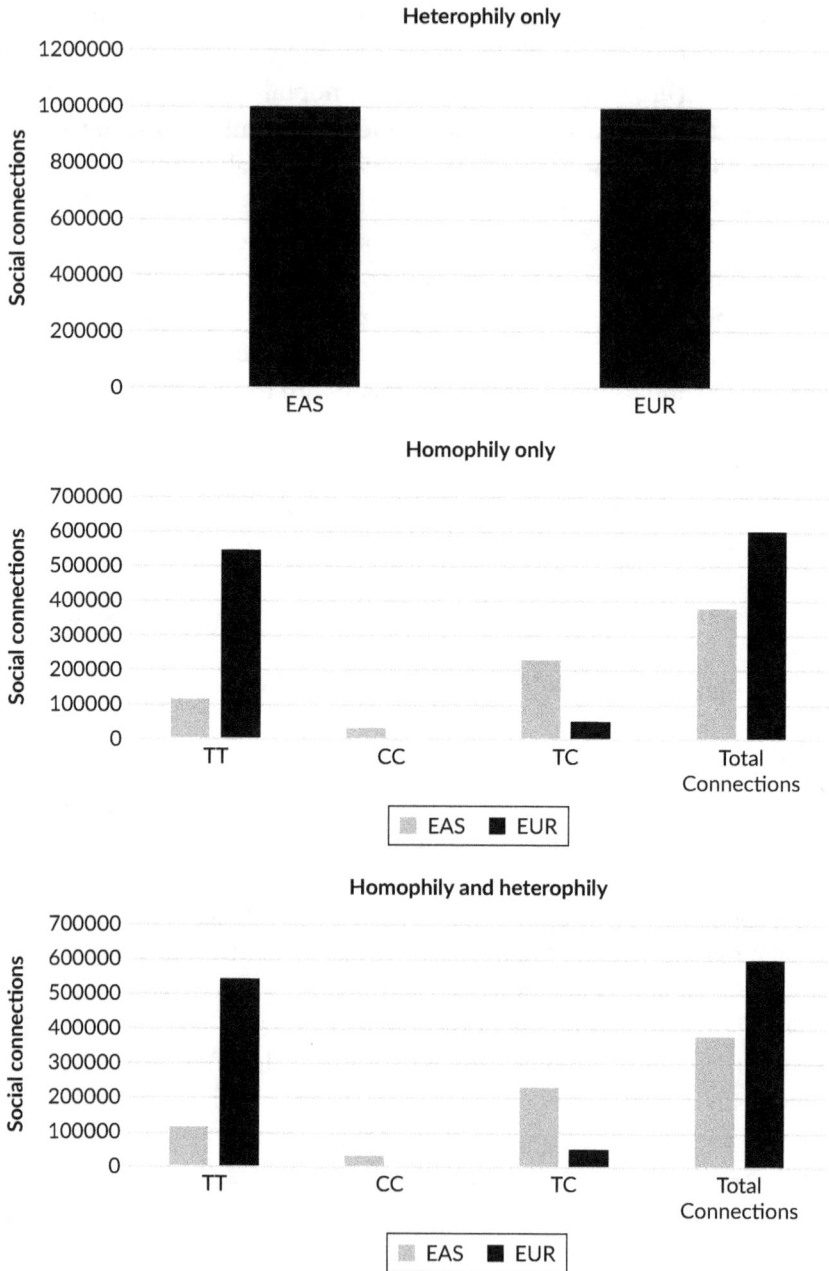

Figure 14. Number of social connections in relation to the genotypes of genes contributing to homophily and heterophily.

simulated by the team's behavior before the cold showers when the authority of the coach was above any doubt, links between team members were minimal, and the whole structure operated in a manner that relied on overall acceptance of each individual's roles beyond questioning. After the cold showers, the team becomes "westernized," relationships between individuals become more tight, complex, and abundant. Also, the authority of the coach, their leader, is under question.

Of course, another controversy emerges with this line of thinking: One would have expected that tight bonds between individuals to prevent the "fluidity" in a society making them less resilient to change. Yet, Westerners appear to be more progressive to change than the Easterners which are more prone to the adherence to traditional ideas. On the other hand, of course, tight bonds also facilitate the transmission of novel ideas. It appears that the actual outcome is the cumulative result of frequently conflicting forces, that change overtime and are almost impossible to predict their specific conclusion.

Thinking about the genetic variations in DRD2 and CYP2A6 in EAS and EUR populations which of the two polymorphisms, among the several that may also exist showing analogous trends, contribute mostly, cannot be inferred. However, the fact that they synergize on the development of a society in the EAS at which weaker social relationships between individuals contribute to its stability strongly suggests that there is a genetic impact on how these societies are organized differently.

PART III

Perspectives

CHAPTER 12

On the Fluctuations and Oscillations of Behavioral Trends

Regardless of the exact mechanism that operated during these extremely large-scale processes, which involved large groups of people and whole populations, various interesting predictions and some testable hypotheses can be made, based on the aforementioned notions. Let's take as an example the *s* and *l* alleles of DRD4 that are associated with collectivistic and individualistic behavior. Since this behavioral pattern appears to play a major role in the distinction between Eastern and Western cultures, it represents a good paradigm in order to attempt a speculation on possible future trends. Analogous considerations can also be made for all other polymorphic genes described earlier, especially if we view them according to the concept of being "en bloc."

From time to time, certain epidemics have occurred in different regions of the world, such as the plague of Athens around 430BC. According to a recent DNA analysis, it was attributed to typhoid fever (Papagrigorakis et al., 2005). Consistent with our previous discussion on the protective consequences and advantages offered by the collectivistic behavior, we may postulate that such epidemics operated as bottlenecks that have enriched for the *s* allele in the corresponding populations, a fact that in turn may have boosted collectivistic behaviors (or vice versa, since collectivistic behavior offered an advantage, it favored the *s* allele-bearing individuals).

Theoretically, the extent of such bottlenecks could have been recorded and calculated. If we were able to obtain DNA from representative groups of individuals dated prior and after such catastrophic events, and assess the frequency of the *s* and *l* alleles, we might have been able to monitor those sequential increases and decreases in the *s* allele's frequency, that transiently favored collectivistic behavior in otherwise individualistic societies:[28]

Each time such an epidemic had occurred, it is conceivable that among the people that died, the ones with the *l* allele are more likely to do so, since they are devoid of the beneficial and protective effects of the *s* allele. At least if we view it at the level of the group and ignore the fact that collectivism is offered by the s allele (or others if acting similarly) but is acting upon both the *s* and the *l* allele individuals. Or, alternatively—and more likely, actually—the ones with the *s* allele, since they were able to adapt more efficiently into the new conditions during the crisis or the ones that followed the crisis, were likely able to produce more offspring. So, it is anticipated that following completion of such a cycle of the epidemic, the *s* allele's frequency would have increased and, probably, the higher the increase, the stronger the selective pressure and, ultimately, the higher the benefits of the collectivistic behavior in this particular case since progressively the society would become more collectivistic. This applies to all genes that contribute to the establishment of a collectivistic behavior.

Apparently, that crises induce collectivism in individuals is something we have all experienced beyond genetics. In the end, the individuals that can do this more efficiently, namely alternate from individualism to collectivism, will become more successful. And by extrapolation, the populations at which the change between collectivistic and individualistic tendencies is more efficient will also be more successful. This reflects the fundamental knowledge according to which survival is quite strongly associated with adaptability and plasticity. To that end, we may understand the presence of relatively high genetic variation among a given society, and the presence of available polymorphic alleles as a safety valve that warrants survival and prosperity in a given society.

In an opposite scenario, in people of collectivistic cultures, we may expect that prolonged periods of "prosperous" living, at the level of several generations, without disastrous phenomena (not only natural) of any kind that are associated with crises that otherwise could have favored collectivistic behaviors and restrain individualism, might stimulate the enrichment of the *l* alleles, at the expense of the *s* allele's frequency. Whether in that case the *l*-allele-bearing individuals (and the corresponding behavior that is associated with that) would have been offered an apparent advantage or rather the advantage of the *s* allele "collectivistic" individuals would have been diminished, giving the chance for phenomena associated with random genetic drift to prevail, is unclear. Anyhow, nowadays the reduction of the prevalence of such disastrous phenomena, and especially epidemics, is more likely to occur and is also favored and facilitated by the advancement of science, technology, and medicine that reduce the negative impact of these events. Thus, under these conditions of prosperity and harmonious co-existence of people during periods that are devoid to major catastrophic events, individualistic behavior is favored, and this may result in the enrichment of the allelic frequency of the *l* alleles.

Historical experience during the twentieth century showed clearly that totalitarian political regimes—which can be viewed, in some extreme sense, as the ill reflections of collectivism—are quite likely to occur after prolonged periods of depression, conditions that induce fear, and share certain similarities with periods at which epidemics and catastrophes may have occurred. This does not necessarily mean that totalitarianism should be strictly related to the communist systems that appeared in Europe, especially during the second half of the twentieth century. Those systems actually claimed that, for them, the development of collectivism represents an essential purpose and the means for a population's prosperity. However, even in societies and in economic-political systems structured on the basis of Western-type individualism, the main elements of such collectivism can be identified. Such an example can be recognized in the emergence of Nazism in Germany that followed the Depression just after the First World War. In that case, the benefits of collectivism for the individual people were found in the inclusion in a political party, a fact that in turn provided a socio-political and financial niche. In any case, following some prolonged crisis, people appear to re-discover the benefits of collectivistic behavior. On a more contemporary example we seem to appreciate the advantages of collectivistic behaviors during the COVID-19 pandemic, at which our dependence on the attitudes of the others is highly enhanced and appreciated.

Naturally, as someone may have well anticipated, harmonious living is not likely to occur forever, especially in groups of people that are devoid of collectivistic norms. Conflicts of interest are intrinsic to societies that are prone, especially to individualistic behaviors and at which the well-being of a single person has been dissociated from (and occasionally opposes) the well-being of the others. The fact that Western societies desperately try to build a collectivistic conscience advocating the values and benefits of belonging into a greater group of people does not contradict but rather reinforces and supports their individualistic orientation (and predisposition). Caring for the neighbors and generally for others, by being active members

157

of the society, and by supporting the ones in need among the group of people that we feel we belong to, likely reflects the fact that such collectivistic notions are imperative in order to be able to function as a single social entity. To that end, such collectivistic notions operate as the cohesive bonds of our societies and are constantly re-enforced by the society through the governments. Of course, we should never forget that nobody is absolutely individualistic or collectivistic at the same time, and this is applicable at an even larger degree to whole groups of people composed of individuals prone to either individualistic or collectivistic behavior. Besides the requirement to generate the appropriate norms and bonds that will facilitate our function as a society, each individual also has the need to satisfy his collectivistic instincts. Thus, a collectivistic mentality is being progressively built, without, of course presenting an imminent danger to the dominant individualistic norms of the society. The opposite, of course, applies to the collectivistic societies that apparently are not devoid of individualism. It is just a matter of the cumulative balance and towards where it is shifted, in either case.

Crises, however, ignited either extrinsically or intrinsically, are eventually unavoidable, and in turn, the socio-environmental conditions at which collectivism will again flourish are unavoidable. An attempt to diagrammatically present this is shown in Figure 15. This is an oversimplified summary of the ideas presented here.

We pose that collectivistic or individualistic behavior is not a straight line, but rather a wave-like line at which peaks are followed by troughs, and recession periods are followed by recovery periods. The vertical axis represents the "strength" of the individualistic or collectivistic behavior that, in turn, can be translated in allelic frequencies in the corresponding populations.

According to this very simple hypothetical diagram, everything started again in Africa, about a hundred thousand years ago, when originally these polymorphisms, the s and l alleles in particular, had already been generated and stabilized in specific subpopulations. Then, small groups of people started migrating out of Africa, already having different frequencies in the corresponding alleles. Then, these alleles either because of the advantages they offered, or because of the genetic drift, or even more likely because of a combination of both, were stabilized in the resulting populations that eventually occupied different regions of the world.

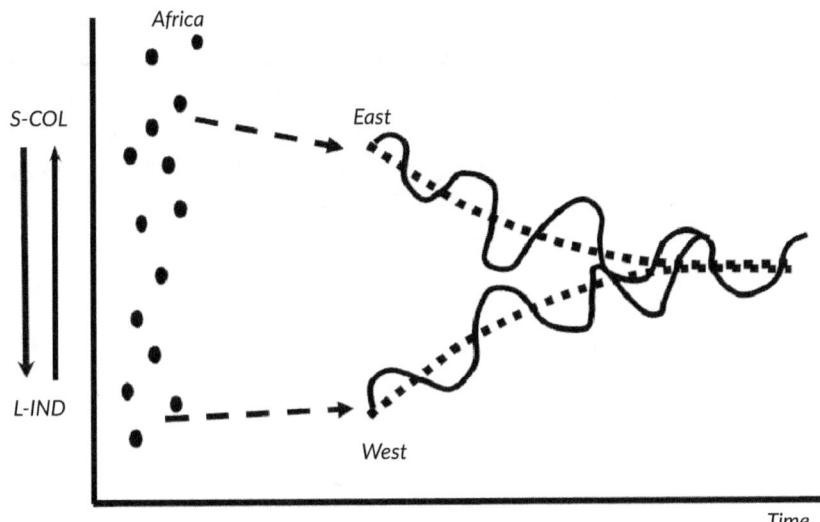

Figure 15. The diagram shows a possible trend for the allelic frequency of *s* and collectivistic behavior (COL), or the allelic frequency of *l* and individualistic behavior in the East and the West. The *x* axis indicates historical time while the vertical *y* axis the degree of collectivism/individualism that is presumably analogous to the frequencies of *s*/*l* alleles in the corresponding populations.

Subsequently, depending on the potential advantage they did (or not) offer to these populations, they started becoming enriched (or replenished), while the corresponding people that "happened" to occupy Eastern Asia or Europe started developing cultures that exhibited the earliest evidence of collectivistic or individualistic behavior, as we know it and record it today.

Naturally, the corresponding allelic frequencies did not remain stable but changed over time. This change could have been due to random events, responses to external effects and changing conditions, or migration from and more importantly towards these populations. We may imagine, though, that each individualistic peak was, and actually still is and will be, followed by an eventual recession that occasionally is signified by a type of catastrophic event of sufficient power and magnitude to ignite a boost in the collectivistic behavior. We can also speculate that if such periods are quite prolonged and at least greater than the duration of a single generation, then they will also dictate analogous changes in the *s* and *l* allelic frequencies that, in turn, will affect behavior even more. Such fluctuations are very common in real life and can be seen in the natural physical phenomena, such as the temperature and CO_2 emissions, to blood pressure and animal populations, and even to the stock market. That the projection of these lines exhibits a decreasing or increasing trend for Easterners or Westerners will be discussed in the following chapter.

CHAPTER 13

Trends

Given the role of *s* and *l* alleles in regulating peoples' behaviors related to an individualistic or collectivistic tendency, a reasonable question is on whether we can predict a future trend that involves these perceptions of life and society. The answer to that question is no. These phenomena are too complex and multi-factorial to be predicted at some level of acceptable scientific accuracy. The intrinsic and extrinsic variables are too many to calculate their contribution in the specific outcome that we are attempting to quantify and interpret. There are, however, a few specific parameters that we may consider as quite likely to occur, and therefore make an "educated guess" regarding possible trends.

First of all, isolated populations are less and less likely to exist, not only physically, but also culturally, economically, and intellectually. We frequently call this trend *globalization*, and as a state it has many proponents and opponents. No matter, however, if this is good or bad, the fact of the matter is that it is unavoidable. It's happening more and more intensively, and the overall "size" of the world in which the future generations are going to live will be smaller than the one we live in today. By that we mean that people tend to live in a single global community instead of several, relatively isolated sub-communities. We can understand its consequences in all the attributes of our everyday life. At the cultural and probably intellectual level, turning the television on is a simple action that is sufficient, though, to show how people live and think at the other end of the world. And this doesn't remain as purely academic information, but has immediate consequences for the everyday lives of people. The effect of American television series in Indian women represents a quite characteristic example: Where various attempts to liberate women, by the introduction of specific laws and regulations, failed, Western television serials have succeeded (Levitt & Dubner, 2009). And this, in turn, had significant consequences on the structure of Indian society. Countless analogous examples can be described at this point on the effects of television all over the world.

Whether we, as inhabitants of a specific geographical area, live better or worse than other people, either closely or remotely located to us, today it is easy and straightforward to find out. And while with the television we can only receive information, with the revolution of the Internet we can also interact (culturally and financially) with these remotely located people. So now, the norms of our society are not being established only by our family, our schools, the formal educational system, and in general the social environment that is physically in our close vicinity and at which we belong, but also by people that can be located physically far away from us. It's the global village idea (McLuhan, 1962), which makes iPads quite compatible with the chadors, and the sushi bars an intrinsic component of the basements of several multinational corporations (and in which Eastern Asians may not necessarily work). Our cultural environment is not defined only by what we have inherited from our ancestors combined with our own contributions to a defined society (that contains us), but also by inputs we continue to receive from remote cultural locations.

Naturally, such external influences by other cultures have always represented a characteristic that occurred constantly during the history of all cultures. In the past, it was the merchants, the priests, the sailors, the soldiers, or other individuals that usually, because of their jobs and missions, were able to interact with populations far away from their home countries and who had received or transmitted cultural elements. The difference now is that the flow of information is not sporadic and occasional, but rather continuous and bidirectional, and cannot be reduced to specific groups of people that, as travelers, operate as the carriers of this information. Following the "discovery" of China by the West, an extensive exchange of goods and cultural attributes was initiated. However, the corresponding cultures and societies were well defined, while the "self-sufficiency" of the corresponding cultures was not at a major risk. What was "foreign" could be defined quite accurately and the exchange was limited to specific merchants, missionaries, and other defined groups of people. What products or ideas that were to be exchanged were not only limited quantitatively, but were also and unavoidably "filtered" before they reached the individuals who were receiving it. So, its effects were relatively limited, specific and somewhat "controllable". As opposed to that pattern, today's flow is continuous, multidirectional, and uncontrollable.

162

In financial issues, the inter-dependability of the world's economies is even more apparent. The financial crisis in Greece around 2010 has attracted a lot of attention from the international media, the world's leaders, and ordinary people all over the world as well. This is not simply because the rest of the world suddenly started to worry about the quality of life of the Greeks, which was becoming worse, but rather because of the fear of the ignition of a domino of collapse in the world's economies. The global consequences of the bankruptcy of the Lehman Brothers investment bank in 2008 provides a good recent example of this global interconnection and interdependence. It's the butterfly effect,[29] according to which the flapping of the wings of a butterfly at the one end of the world can cause a hurricane at the other end of the world!

Another emerging issue of modern times is the progressive reduction of physical isolation among people of different regions of the world. Migration always existed throughout the history of mankind, and therefore the exchange of genes (alleles) was happening. Since no populations (of considerable size) were absolutely isolated for considerable time, gene flow had always occurred.[30] Mostly, those exchanges involved historically defined movements of groups of people of which their relocation was tightly related to specific natural phenomena and needs within a given historical context. It could have been famine, political oppression, slave trading, or specific natural phenomena (disasters) that forced, motivated, or ultimately persuaded people to relocate and seek a new homeland. Moreover, they refer to a defined group of settlers that either occupied a previously unoccupied region or entered a region that was occupied by another, local population with which the migrating population eventually intermixed. The genetic pools, though, were relatively well defined, despite the fact that these pools were never isolated. Intermixing did occur but at more reduced rates, especially when this involved genetically "distal" populations. In other words, the "moves" (of people and genes) were tractable and historically recordable. And even in the cases in which more extensive intermixing occurred, the resulting population (as well as the populations from which the latter originated) was becoming again quite well defined and distinct. Take as an example the massive waves of Spanish and Portuguese (along with their slaves) that inhabited South America. They interbred with the indigenous populations and the resulting, contemporary inhabitants of South America can trace their ancestors, to some extent, to these original populations (notwithstanding the fact that frequently and quite apparently the one's genetic signature predominates over the other's in different people).

Today, though, assisted by the increasing cultural interconnectivity, the exchange of genes is becoming easier between individuals of different populations and the genetic flow more unbiased and smoother. Thus, the global genetic pool tends to be unified, or in other words, to become a single pool with different densities in the allelic frequencies in particular regions. Within this greater population, the flow of genes will be more or less continuous (Figure 16).

It is likely that allelic frequencies will tend to vary relatively smoothly with location, while cultural and other barriers will be less strict.[31] Moving and relocation of people from one place of the world towards another is becoming cheaper, more and more easy, feasible, and applicable. Of course, the trends are not the same and equivalent between opposite directions of such potential moves. At least during the current, specific historical period there are certainly more Chinese that want to relocate to the U.S.A. than Americans to China, or Pakistanis to Europe, than Europeans to Pakistan. But the fact is that continuous waves of immigration do occur that ultimately tend to result in some homogenization of the genetic pools. And this trend of homogenization will not only occur at the genetic level, but initially tends to occur at the cultural level, as well. At first, these immigrants tend to live in relatively isolated sub-communities in which genetic

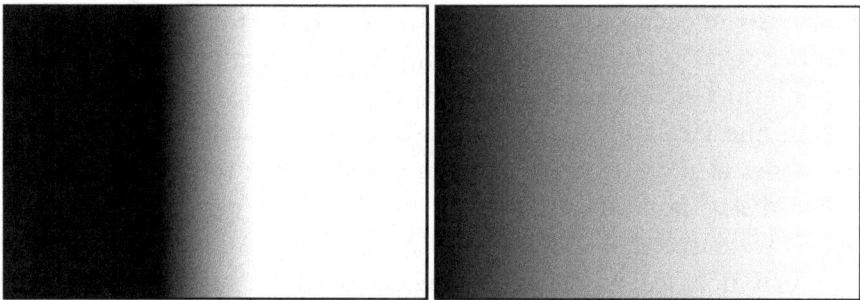

Figure 16. A schematic diagram depicting the difference between sharp (left) and smooth (right) changes in allelic frequencies (density of black or white color) in two populations at which genetic exchanges have occurred. In the left, the populations are relatively still well defined. In the right, the populations tend toward homogeneity. Presumably, the current global tendency is towards the right state.

164

exchanges with outsiders are limited. Subsequently, though, after a few generations, assimilation is complete and genetic flow will likely be unrestricted. No matter how massive this trend will be, phenomena related to the stratification of the populations progressively will tend to have reduced impact.

Last, but not least, another characteristic of modern times is the progress in science, technology, and medicine that tends to reduce the impact of certain physical disasters and diseases. This becomes quite apparent as regards infectious diseases for which the overall improvement in the quality of life and in sanitary practices, combined with the use of antibiotics and other potent medicines for which the access is becoming wider, renders the consequences of these diseases in the population less devastating, as compared to what was happening during the past times.

Of course, certain regions of the world, particularly in Africa, do not benefit from these advancements as much as other areas in the rest of the world, at least and hopefully not as yet. However, it is rather likely that given the increasing sensitization of the people around the globe, eventually these benefits will pass even to the people of Africa who, at present, seem to receive and benefit from it, the smallest—if any at all—fraction of humankind's scientific and technological progress.

Talking about conditions that endanger certain populations, it has to be noted that contrary to the beneficial consequences of science, technology, and medicine, a clear danger persists that has a magnitude that is by far greater than analogous dangers met by humanity before: It is related to the impact that wars may have today and are likely to have in the future. This is due to their capacity to be much more massive and devastating, as compared to what mankind ever dealt with before. Indeed, while in past encounters the casualties were limited to thousands or tens of thousands (solders in particular), today the magnitude may be increased to the level of millions and even more.

Collectively, the fact that death from "unnatural causes," as we use to call them, becomes less and less likely in the days to come, the conditions that boost collective behaviors are going to be less common and frequent. In turn, and at the first instance, this reduces the advantages linked to these behaviors. Thus, the bottlenecks we speculated earlier are going to be less drastic and potent. Yet, while the benefits of collectivism in terms of coping with epidemics will not be that demanding, the requirement for social support due to other psychosomatic factors related to loneliness, for example, will be of need. Furthermore, this will be needed more according to the studies of individuals that have the genetic makeup of collectivism but now live in individualistic societies. This is reinforced even more by the fact that they are usually immigrants in which the weight of stress due to their need to succeed is even heavier. Sociological approaches studying the integration of immigrants may likely have to consider all these in the future and attempt to resolve potential issues by considering the genetic parameter as a contributor as well.

Keeping all these in mind, and for as long as the Easterner vs. Westerner cultural distinction continues to exist, we may be able to identify a trend for Easterners towards a more individualistic pattern. Of course, in that case, we have followed the "reverse" approach, according to which it's not the genes that dictate behavior but rather the socio-cultural (world's) environment that favors the specific behaviors that, in turn, may offer a long-term advantage to the individuals with specific genotypes. We can already see this at the cultural level, when we identify an increasing evidence for the invasion of elements from Western culture in the East, and it appears that this is much deeper than just a cultural trend. As already mentioned, and at the extent at which the present socio-environmental conditions no longer favor—at least not that much (!)—collectivistic behaviors, the latter become more vulnerable to individualistic attitudes. So, eventually they will be replaced by patterns and norms that are more suited to l-allele individuals. So, in Easterners, in the graph at Figure 15, we may be able to identify a tendency for the reduction in the frequency of the s allele and the associated collectivistic behavior in the long run.

But what happens with the already-hardwired-for-individualism Westerners? It is conceivable that the predominant parameters that will govern their future and fate are the following: The gradual infiltration of Western culture by Eastern values and norms may pull Westerners towards some collectivism. This is reinforced by the fact that currently, and likely for the years to come, Asians that immigrate towards the West are more, arithmetically speaking, than those exhibiting the opposite pattern of immigration, from the West towards the East. So the overall balance in such Western populations shifts towards some collectivism, at least in part due to their enrichment for "Eastern" s alleles. In addition to that, no matter how strong the efforts to transcend a collectivistic behavior in the people are, as it happens currently in the West, in order to preserve the ability of a group of people to function as a society, I doubt that this will ever be sufficient. Political, financial, and social crises that are intrinsically associated to the extreme individualism are unavoidable, and those cannot inhibit the occasional emergence of conditions that, in turn, will favor collectivism. And these crises may be even stronger with time, since sustainable development, by Western standards, is not that feasible to achieve continuously in the future.

The increasing attraction of Western people for Eastern (and frequently collectivistic) ideas, the surge of community values, and even the occasionally extreme shifts towards certain religious groups, are much more than just a superficial trend. Probably, they reflect inherent attempts to satisfy Westerners' suppressed collectivistic needs that are becoming lost in a flourishing individualistic and complex society. And the greater and more interconnected these societies are, the higher the demand for the rediscovery of collectivism will be. Thus, we can predict that Westerners will gradually shift towards some collectivism at some degree. Again, this shift will not be straight and acute, but rather it will be the cumulative outcome of successive cycles of collectivism and individualism. Westerners will have to learn how to live more collectivistically, otherwise they will not be able to function as a unified society. Analogous to that is the conclusion of Tim Jackson (2011), who, based on an economic analysis, argued that the continuous reinforcement of individualism is problematic and a shift towards certain social values should be pursued.

13.1. Can Novel Cultural Trends Emerge During Globalization?

According to Figure 15, the shift towards collectivism and the conceivable increase in the frequency of the s allele in the long run in Westerners was accompanied by an analogous but opposite shift of Easterners towards individualism and a reduction in the frequency of the s allele. Naturally, this is rather hypothetical and by no means indicative for that the "merged" population will simply be the direct and proportional average of the two populations. It is not only that many other genes can affect the state of the final balance, but also that it is very naïve to predict that what is "lost" in terms of collectivism from one population will be gained by the other population. In other words, I doubt that the "homogenized single population" will function that democratically that its constituent subpopulations will contribute equally.

167

If we go back again to Figure 15 and given the trend for homogenization of the various cultures and populations in the world, we will be able to predict that the ultimate outcome will be to obtain a heterogeneous but single population that will also show evidence of fluctuations and oscillations. In other words, it is also quite accurate to predict that the high interconnectivity of the various populations that exist around the world will make them able to fluctuate or oscillate together and not independently. This notion of homogeneity, as an ultimate stage of societal organization, has been advocated by various political analysts and philosophers, with Francis Fukuyama (1993) more recently, being one who predicted the *end of history* in his classic and controversial work. In his work, Fukuyama has envisaged the Western-type liberal democracy as the ultimate political state, but again the predominant notion is that of homogeneity that will transcend political administration, financial relationships, and cultures.[32]

So, certain periods during which collectivistic behavior will appear as having the predominant role in society will generate unavoidably conditions that will eventually favor individualistic patterns and norms. In turn, these individualistic patterns and attitudes will generate crises that will be resolved by the advancement of collectivistic behaviors. And this will go on until some quite unpredictable and massive event disturbs this equilibrium and pushes this dynamic balance strongly towards the one or the other direction. It has to be noted that these transitions from the individualistic toward collectivistic tendencies will not be guided only by causative changes and selection processes. Other phenomena related to genetic drift and population stratification may also be able to cause analogous changes; however, it is plausible that the signifying issue is that the whole world's population will tend to behave as more unified as compared to how it behaved in the past.

An alternative, intriguing possibility may also happen. Obviously, it is more than a single gene, besides DRD4, that regulates collectivism and individualism. We already saw that groups of polymorphisms have coordinated frequencies worldwide, differ in East Asians and European Ancestry and all together are implicated in the development of their cultural trends. While the Asians shift to individualism and the Europeans to collectivism, to do that various possibilities in allelic combinations exist beyond the simple reversal of the allelic frequencies in the corresponding populations. For example, let's assume that the frequency of genes A, B, and C are high in Asians and low in Europeans. Instead of having the frequencies of these genes A, B and C to increase proportionally in Europeans and decrease in Asians, it is possible that A will increase disproportionally in the Europeans and C will decrease disproportionally in Asians. Or another gene D may increase or decrease in one population, causing similar effects. Such changes, while they may satisfy the need for individualism and collectivism respectively, in the long run they also provide the genetic background for novel cultural trends to emerge leading to developments that ultimately reflect the development of cultures that are qualitatively distinct from the ones we know today. And what will trigger this will be the ever-increasing pressure to different populations to operate under norms that were not accustomed or hardwired for them. In that case we may witness novel cultural identities emerging that deviate from the individualistic-collectivistic dichotomy that we know today.

It is also quite likely that during these processes, some relatively isolated societies will also be generated, or they will continue to exist within the greater global society at which the consequences of this homogenization will have a smaller effect and will be less apparent. The impact, though, of these societies in the global population will likely be relatively small. The populations of the gypsies (Roma) around Europe or that of the Amish people in the U.S.A., offer only some of the many examples of subpopulations of people that despite that they live physically within the borders of a greater society, their cultural and more importantly genetic exchanges with outside people are restricted. These relatively distinct populations and, until the time (and of course, if) they become absorbed by other populations, will continue to function independently; however, the fact that they represent only a minor fraction of the world's population reduces their impact in the global cultural trends.

Related to that is also the fact that "globalization" does not progress at the same rate around the world. Countries like the U.S.A. continuously accumulate a tremendous amount of genetic variation in their population through immigration. It is likely that in a few generations, the genetic diversity of people in the United States will approach that of Africa, where all different alleles can be found and eventually intermix.

So, according to Figure 15, we can predict that in the eventual single "homogenized" populations, cycles of reduction followed by increases in the allelic frequencies of the s alleles will periodically occur. For how long these cycles will last, it is obviously hard to predict, but in order to have

169

some balance restored between the peoples' intrinsically dictated attitudes and their cultural and societal norms, by definition they will need to go on for more than a single generation. Otherwise, it is hard to imagine that it will be the same group of people, having the same exact frequencies in their corresponding pool of alleles, which while at first, they exhibited a trend for individualistic behavior, they then continued by exhibiting collectivistic patterns and attitudes. Genetic time, especially at the level of populations is by far greater than historical time. It is counted in generations and not in years. Certain changes in the allelic frequencies need to occur, a process that, by definition, and in order to reach a form of stability or equilibrium, requires more than a single generation. This way, the actual "genetic predisposition" and the occurring social structure will maintain a form of balance with each other.

Analogous mechanisms may operate regarding other behavioral traits that exhibit differential tendency, due to certain genetic frequencies, between East and West. Again, such as in the case of the previously discussed *s* and *l* alleles, the "homogenization" of global society will result in the generation of a single world population that will display evidences of fluctuations and oscillations regarding the allelic frequencies and the corresponding behavioral patterns. The idea is that the increasing efficiency and benefit offered by the one allele of a given polymorphism, at some point will generate conditions that will inhibit its further increase, a point at which the individuals (and the corresponding subpopulations) with the alternative allele will start exhibiting higher benefit and have higher advantage. In turn, at these conditions, this other allele is likely to increase in frequency, generating the cycles of periodic oscillations and fluctuations that eventually will lead to a more homogenous global population with the corresponding overall allelic frequencies in between the original values.

CHAPTER 14

Isaac Asimov's Psychohistory and the Prediction of History

Isaac Asimov, in his *Foundation* series of novels, has introduced the speculative science of psychohistory, a science that by combining history, sociology, and mathematical statistics can make general predictions of future behaviors of very large groups of people. In Asimov's hypothetical universe, with the help of psychohistory, people were able to overcome certain crises following the advice of a mathematician by training named Hari Seldon, psychohistory's inventor, developer, and actual practitioner, who appears at specific times at which certain vital decisions should be made. By being able to apply psychohistory's axioms, he was able to predict history and help people render the correct choices and decisions. The success of psychohistorical predictions is heavily based, according to Asimov's Seldon, on the involvement of large populations that have complete ignorance of the results of the psychohistorical analyses. Only under these conditions is psychohistory able to make accurate and valid predictions.

Equally interesting is that in this hypothetical universe, the original settling of selected colonies destined to maintain humankind and intellect was performed by a selected group of people with specific abilities and skills. Furthermore, within this universe, the author describes two major historical lines (or cultures) of humankind, one based on rational, logic, and scientific advancement (First Foundation), and the other based on individuals with mentalic abilities, telepathy and the ability to adjust human emotions (Second Foundation).

In subsequent books that also take place in this virtual universe in which psychohistory is operational, Asimov also implies that the Second Foundation represents the embryonic stage of the development of a collective consciousness, since individuality in this world is not strong and gives way to the needs of the greater whole. The ultimate stage of such collectivism is a complete unification of all elements of the world to a single super-collective entity, exemplified by the concept of *Gaia*. Collectivism, according to Asimov, is not related to logic and scientific advancement.

An imminent similarity with the two actual cultural lines of humanity, of the West and the East, can be identified that apparently inspired Asimov. The rational, individualistic, and science-prone First Foundation can trace its roots to the Western line of thought, whereas the collectivistic Second Foundation to the Eastern line of thought.

Since we know now that at least a fraction (and likely more than that) of our behavior is due to our genes, renders theoretically feasible the development of a scientific discipline, analogous to psychohistory, that would be able to gather and analyze large-scale data and make certain predictions. Interestingly, we can also imagine that in order to render predictions that will be as accurate as possible, data have to involve huge groups of people and also that the individuals constituting these groups should not be aware of the results of this analysis, and thus adjust accordingly their behavior. Otherwise, the results would be biased and inaccurate.[33]

Despite the many and various limitations of this concept, of which only a small subset has been described earlier, some hints regarding future trends can be made. Naturally, all these might be applicable only in cases at which dramatic changes in the genetic composition of the existing populations would not have occurred (see chapters 12 and 13 for details). Otherwise, if populations became unified as regards to their gene frequencies, different behavioral patterns could not have been attributed to a genetic factor. In other words, these may apply prior to the conceivable, previously described "homogenization" of cultures and populations.

By keeping this limitation in our minds, we may be able to foresee certain tendencies. For example, given the advantages of the collectivistic behavior during periods of crises and disasters, we can predict that populations with a tendency to collectivism at such times may exhibit an advantage and higher efficiency. Assuming, for example, that such an event or series of concomitant events have occurred at a global level, it is quite likely that Easterners, that have an inherent tendency for collectivism, would have a certain advantage in terms of being able to function better as a society, or as a whole. During the COVID-19 pandemic, we experienced the effectiveness of China in containing the pandemic, compared to Western countries. This has a certain impact on the economy, with China's continuing to grow, as opposed to the U.S. economy that is expected to regain growth only in 2021 or even later. This offers a competitive advantage to China.[34]

The type and the extent of such disastrous events are quite subjective. For example, what for a given society is considered as a regular, though negative, event, in another society may be disastrous. This of course doesn't have to do with the value of individual human life, but rather with how the consequences of such events affected and actually altered the regular functions of the corresponding societies. And this doesn't always have to do with different mentalities, but also to how accustomed the populations are to specific events and eventually adapt and adjust accordingly, to the new norms that are mandated. Take as another example the consequences of hurricanes in cities around the Gulf of Mexico, and of earthquakes in Greece. These are geographical regions that are familiar to such natural disasters, and imagine what would have happened if such a hurricane hit Athens. The residents however of Greece, view earthquakes as an event that is not that uncommon as compared to hurricanes, they built houses that can better sustain them and more importantly, as a society, they resume functioning soon after such events happen. Vice versa, the impact of a strong earthquake might have been much more devastating and scarier, to the residents of Louisiana that are used to hurricanes but not to earthquakes.

Another example that actually involves populations with different attitudes is related to the way that the Japanese society and state dealt with the recent Fukushima disaster.[35] Try to envisage what would have happened if an analogous event of similar magnitude hit California. In all cases, the examples referred to wealthy Western or Westernized societies, at which the direct consequences of the disasters, at the first instance at least, involved only specific communities and populations. Thus, only indirectly did they have global consequences, but certainly they had imminent effects in the populations that were directly involved.

If we try to imagine the consequences of events of global magnitude, the effects might be much greater. Such events may not necessarily refer to physical and unpredictable sudden disasters, but with intrinsic crises of the current societies. Such crises can be financial, political, or even military. Take, as an example, the financial crises of 2010 in Europe that had a domino effect on economies around the world.[36] If the economical crises of the last years became more intense, persistent, and global, the collectivistic populations might exhibit more efficient approaches and strategies to cope with the potential damages. Such a scenario, on the likelihood of

173

a forthcoming major crisis, according to many economists and political analysts is not that hypothetical and might even be unavoidable.

Analogous and even more apparent may be the consequences of generalized war-like situations and military conflicts at which (and afterwards usually) collectivistic behaviors have certain advantages to exhibit. It is not difficult to imagine that for individualistic societies, socio-economical deterioration is more possible under tense poverty and damaging conditions than for collectivistic societies. The former societal and political structures are more vulnerable than the collectivistic ones because, at some point during the crises, they may attempt to resolute the conflicts and the disputes by changing their intrinsic norms and mode of operation. And while at some point this could be considered as adaptability, after that, such changes may lead to instability and disorganization.

On the other hand, for as long as these intense situations and crises at which collectivism is favored do not occur, societies in which the majority of the population is "hardwired" for individualism may have an advantage. Thus, such societies may be more efficient and progressive.

Intuitively speaking, it seems that individualism is good for advancement and development while collectivism for restoration of balances after the occurrence of crises—and eventually for damage control! The latter may also be reflected at present times of perceived generalized stability and peace, to the fact that Western-type norms appear very appealing worldwide, despite that at the first instance, the cultural elements they introduce at the various other cultures around the world appear irrelevant to the peoples' cultural heritage. Vice versa, the frequent re-discovery of Buddhism by celebrities of Western society that appear very successful within their individualistic societies may be relevant to this!

Such consequences of individualism vs. collectivism can be seen at other levels, as well, which do not necessarily involve imminent crises. At the political level, for example, historically liberal societies that have the Westerners' presumed genetic imprint of the high frequency 7-repeat allele for DRD4 (liberal behavior as described previously)—and who keep on accepting as residents, people with lower frequency of this "liberal" allele— may start exhibiting a tendency for conservatism. It is noted at this point that liberalism and conservatism should not be limited to the traditional notions of the current political spectra. In other words, one may be quite conservative in his life and his ideas despite voting for a liberal political party and vice versa, voting for the conservatives but introducing and practicing quite revolutionary ideas. With these issues in mind, at the same time, and given the "en bloc" concept of these tendencies described earlier, such societies may eventually start exhibiting an increased trend of altruism that in its essence is not fully compatible with such societies' competitive and individualistic norms. At the same time, the complex, Western-structured societies at which the "*worriers*" predominate will have individuals with "warrior"-type behavior increase in numbers. Presumably those, and according to their reduced efficiency under complex socio-economical environments, may shift the societal norms towards directions that are better suited for "warriors." In that case, we (or our great-grandchildren, more likely) may witness changes in the politico-economical systems not intrinsically related to the development of capitalism and liberal democracy per se that can be explained by pure financial and socio-political criteria, but are rather due to the collective outcome of the changes of both the related financial parameters and the genetic constitution of the individuals practicing it.

Analogous will be the challenges for the populations of the collectivistic societies at which individuals besides collectivism also show tendencies for conservatism, lower novelty-seeking and risk-taking, as well as warrior-type behaviors. The fact that they have to operate at societies that require increased novelty-seeking and risk-taking tendencies and ultimately individualism, may generate certain difficulties. Those eventually will have to be solved by shifting the whole societal structures towards the better suited for them collectivistic direction. Alternatively, they may generate, in turn, certain crises that will have to be resolved by more violent and acute ways.

Numerous such examples can be imagined, all of which have as a common denominator the fact that populations with a given genetic composition—and, ultimately, behavioral tendencies and cultural norms—have to adapt within a limited time period to conditions that appear to be better suited for other populations with different genetic frequencies and behavioral tendencies. Coming back to the crises-prosperity cycles they will ultimately trigger transitions from individualism to collectivism. It depends to the actual populations whether they will implement these changes efficiently.

Of course, what I described here is a snapshot, and what may have sounded problematic is actually not very different from what constantly happened during history, whenever populations with different cultures—and consisting by individuals of likely different genetic frequencies—had to interact, blend, or share materials, customs, and genes. Some particularities due to the widest possible extent of today's and likely future trends have been described in some detail in chapter 12. They are all related to what is making us see our contemporary world different from others, or our times more interesting and unpredictable than the past times. In addition to all those, the time periods at which such changes have to occur are likely shorter than before, and they are more massive than what mankind has witnessed in the past, which makes such current changes quite intense. If I were to single out one parameter that distinguishes the perception of the current versus the older times, I would probably choose this one, being the condensation of the events that make time pass faster.

Epilogue

The challenging question now, by having such scientific knowledge in our hands—namely to envisage and why or why not—is to predict certain behavioral trends in different populations, if we can minimize certain damages and make the resolution of crises smoother. Such notions, of course, if feasible, extend beyond prediction and contain within them hidden the notion of the manipulation of human history.

If prediction is achievable, then it is conceivable that, by altering the parameters of the system, one can change the potential outcome. This is consistent to some type of primitive historical engineering, according to which certain historical choices will take into consideration genetic tendencies in order to make the potential outcome as beneficial as possible for those that can perform such analyses. Indeed, unavoidably, this is the foreseen case that may even sound scary, or that recalls specific memories affiliated with very unfortunate moments of human history. However, I don't think that this differs strikingly from the whole concept and deepest essence of technological and scientific progress, according to which by taking advantage of certain physical laws and technological developments, we aim to intervene with various natural phenomena for the benefit of mankind. After all, this is what human progress is all about throughout the history of our species.

The deepest danger, though, is that such an approach may reduce the value of individual life to the level of a fraction, or just a negligible constituent of a greater whole or specific population. To that end, a clearly fearful outcome is the potential demoralization of certain actions that, presumably for the benefit of mankind (and even worse than that, for the benefit of a group of people or the population), may harm certain individuals.

No matter what, though, the fact is that the genetic identity of a group of people should not be ignored when their history is studied. Besides geography, availability of natural resources, culture, and socio-economical context in general, the behavior of the people, as it is influenced by their genes, should also be taken into consideration. In view of that, the contribution of the various other parameters involved in human history may acquire a significance that is different from what is traditionally thought.

References

Chapter 1

Castoriades Cornelius (1964) in Anthology Socialisme ou Barbarie' - No. 35.

Christakis NA, Fowler JH. Friendship and natural selection. Proc Natl Acad Sci U S A. 2014 Jul 22;111 Suppl 3(Suppl 3):10796–801. doi: 10.1073/pnas.1400825111. Epub 2014 Jul 14. PMID: 25024208; PMCID: PMC4113922.

Fowler JH, Settle JE, Christakis NA (2011). Correlated genotypes in friendship networks. Proc Natl Acad Sci USA 108(5):1993–1997.

Fujimoto, A., Nishida, N., Kimura, R. et al. FGFR2 is associated with hair thickness in Asian populations. J Hum Genet 54, 461–465 (2009). https://doi.org/10.1038/jhg.2009.61

Molnar S. (2005). Human Variation: Races, Types, and Ethnic Groups (6th Edition). Prentice Hall. Upper Saddle River. 416 pages. *A textbook-oriented description of human variation and human races.*

Roberts J.M. (1997). A short history of the world (1st Edition). Oxford University Press. New York. 560 pages. *The classic concise book on the history of mankind.*

Sulem, P., Gudbjartsson, D., Stacey, S. et al. Genetic determinants of hair, eye and skin pigmentation in Europeans. Nat Genet 39, 1443–1452 (2007). https://doi.org/10.1038/ng.2007.13

Chapter 2

Bonné-Tamir B, Nystuen A, Seroussi E, Kalinsky H, Kwitek-Black AE, Korostishevsky M, Adato A, Sheffield VC. (1997). Usher syndrome in the Samaritans: strengths and limitations of using inbred isolated populations to identify genes causing recessive disorders. Am J Phys Anthropol. 104:193–200. *An interesting article describing the genetics of Usher syndrome, their prevalence in Samaritans as well as related anthropological and evolutionary aspects of the disease.*

Cronqvist H and Siegel S. (2011). The Origins of Savings Behavior AFA 2011 Denver Meetings Paper. Available at SSRN: http://ssrn.com/abstract=1649790. *Genetics and savings behavior. The study describes the genetic and environmental contribution to savings behavior.*

Foucault M (2006). History of madness. Routledge, Taylor & Francis. Oxford, UK. 776 pages. *This is the classic work of the French philosopher Michel Foucault that was first published in France in 1961 as "Folie et Déraison: Histoire de la Folie à l'âge Classique".*

Futuyma D. (2009). Evolution. (2nd Edition). Sinauer Associates. Sunderland, MA 545 pages. *A more in-depth textbook on evolution than the previous one that also deals in some higher detail with issues related to population genetics.*

Griffiths AJF, Miller JH, Suzuki DT, Lewontin RC, Gelbart WM (2000). An Introduction to Genetic Analysis. (7th edition). W. H. Freeman and Company, New York. 860 pages. *The classic textbook describing the science and the limitations of DNA analysis.*

Lehrer J (2009). Howe we decided. Houghton Mifflin Co, 302 pages. *An interesting and vivid account of the decision making process.*

Ridley M (2003). Evolution (3rd edition). Wiley-Blackwell. Hoboken, NJ, 792 pages. *An excellent book on evolution that also offers basic knowledge on genetics.*

Chapter 3

Amos H and Hoffman JI. (2009). Evidence that two main bottleneck events shaped modern human genetic diversity. Proc. R. Soc. B doi: 10.1098/rspb.2009.1473. *The genetic analysis that showed that the diversity of modern humans is due to two major bottleneck events, Out-of-Africa and through the Bering Strait.*

Cavalli-Sforza LL and Cavalli-Sforza F. (1996). The Great Human Diasporas: The History Of Diversity and Evolution. Perseus Books. 320 pages.

Cavalli-Sforza LL. (2001). Genes, Peoples, and Languages. University of California Press. 239 pages. *Two of the books of the great human geneticist Cavalli-Sforza that explore the genetic history of human populations.*

de Vaal Malefijt A. *Homo monstrus.* Scientific American (1968) 112–118. *An interesting account of how Homo monstrus was perceived throughtout history.*

Green RE. et al. (2010). A Draft Sequence of the Neandertal Genome. Science 238, 710–722. *In this seminal article the sequence of the Neanderthal genome is reported along with various interesting observations regarding the relation of Neanderthals' DNA with that of modern humans.*

Liu, W., Martinón-Torres, M., Cai, Yj. et al. The earliest unequivocally modern humans in southern China. Nature 526, 696–699 (2015). *Evidence for the presence of human teeth between 80,000 and 120,000 years old in Fuyan Cave in Daoxian in southern China.*

Marks J (1995). Human Biodiversity: Genes, Race, and History (Foundations of Human Behavior). Aldine Transaction. Piscataway, NJ. 321 pages. *Interesting account describing older (including Linneaus') notions about human diversity and races.*

Niewoehner WA. (2001). Behavioral inferences from the SkhulyQafzeh early modern human hand remains. PNAS 98, 2979–2984. *The analysis of the anatomy of the Skhul samples that pointed to the behavioral differences between modern humans and Neanderthals that have already emerged 100,000 years ago.*

Olson S (2002). Mapping Human History: Discovering the Past Through Our Genes. Houghton Mifflin Harcourt. Boston, MA. 304 pages. *Two among the many recent excellent accounts of our genetic history.*

Quach H, Rotival M, Pothlichet J, Loh YE, Dannemann M, Zidane N, Laval G, Patin E, Harmant C, Lopez M, Deschamps M, Naffakh N, Duffy D, Coen A, Leroux-Roels G, Clément F, Boland A, Deleuze JF, Kelso J, Albert ML, Quintana-Murci L. Genetic Adaptation and Neandertal Admixture Shaped the Immune System of Human Populations. Cell. 2016 Oct 20; 167(3):643–656. e17. doi: 10.1016/j.cell.2016.09.024. PMID: 27768888; PMCID: PMC5075285. *The study explores how archaic human DNA influences immune responses.*

Rasmussen M, et al. (2011). An Aboriginal Australian Genome Reveals Separate Human Dispersals into Asia Science DOI: 10.1126/science.1211177. *The study that reports the genome of Australian Aboriginals and their relations to other present day human populations.*

Simonti CN, Vernot B, Bastarache L, Bottinger E, Carrell DS, Chisholm RL, Crosslin DR, Hebbring SJ, Jarvik GP, Kullo IJ, Li R, Pathak J, Ritchie MD, Roden DM, Verma SS, Tromp G, Prato JD, Bush WS, Akey JM, Denny JC, Capra JA. The phenotypic legacy of admixture between modern humans and Neandertals. Science. 2016 Feb 12; 351(6274):737–41. doi: 10.1126/science. aad2149. PMID: 26912863; PMCID: PMC4849557. *This study explores the phenotypic consequences of Neanderthal DNA in modern humans.*

Sykes B (2002). The Seven Daughters of Eve: The Science That Reveals Our Genetic Ancestry. W. W. Norton & Company. New York. 320 pages.

Chapter 5

Lai L.K. An Introduction to Chinese Philosophy (Cambridge Introductions to Philosophy). (2008). Cambridge University Press. 328 pages. *A comprehensive textbook to early Chinese philosophy.*

Nisbett R. The Geography of Thought: How Asians and Westerners Think Differently...and Why. (2003). Free Press. New York, 288 pages. *A very stimulating account of the differences between Eastern and Western people.*

Surowiecki J. The wisdom of crowds (2004). Doubleday; Anchor. 336 pages. *As the title indicates this very well prepared and researched book explores the wisdom and correctness of collective decisions.*

Ferguson N. Civilization (2011). Allen Lane. 432 pages. *The controversial historian's latest book that investigates the routes of Western predomination.*

Chapter 6

McElreath R and Boyd R. Mathematical Models of Social Evolution: A Guide for the Perplexed. (2007). University Of Chicago Press. 425 pages.

Borrello ME. Evolutionary Restraints: The Contentious History of Group Selection. (2010). University Of Chicago Press. 240 pages. *The evolution of social behavior and the selection at the level of a group remains under strong debate in biology. These books introduce critically the reader to the field and the debate.*

Chapter 7

Benjamin *et al.* Population and familial association between the D4 dopamine receptor gene and measures of Novelty Seeking. Nat Genet. 1996 Jan;12(1):81–4.

Ebstein *et al.* Dopamine D4 receptor (D4DR) exon III polymorphism associated with the human personality trait of Novelty Seeking. (1996) Nature Genetics 12, 78–80. *The first genetic association studies linking DRD4 polymorphisms with novelty seeking behavior.*

Paterson *et al.* Dopamine D4 receptor gene: novelty or nonsense? Neuropsychopharmacology. (1999). 21, 3–16. *Critique on the DRD4-novelty seeking association studies*

Wang *et al.* The Genetic Architecture of Selection at the Human Dopamine Receptor D4 (DRD4) Gene Locus. (2004). Am. J. Hum. Genet. 74, 931–944. This article *discusses the genetics and the evolution of the DRD4 polymorphic alleles.*

182

Ding *et al.* Evidence of positive selection acting at the human dopamine receptor D4 gene locus. (2002). PNAS 99, 309–314. *Molecular evidence that the DRD4 7R allele is under positive selection.*

Chang *et al.* The world-wide distribution of allele frequencies at the human dopamine D4 receptor locus. (1996) Hum Genet. 98, 91–101. *A detailed description of the DRD4 allelic frequencies in different ethnic populations.*

Chen C, Burton M, Greenberger E, Dmitrieva J. Population migration and the variation of dopamine D4 receptor (DRD4) allele frequencies around the globe. (1999). Evol Hum Behav 20, 309–324.

Matthews LJ and Butler PM. Novelty-seeking DRD4 polymorphisms are associated with human migration distance out-of-Africa after controlling for neutral population gene structure. Am J Phys Anthropol. (2011) 145, 382–389. *Studies correlating DRD4 alleles with Out-of-Africa migratory geographical distances.*

Eisenberg DTA, Campbell B, Gray PB, Sorenson MD. Dopamine receptor genetic polymorphisms and body composition in undernourished pastoralists: An exploration of nutrition indices among nomadic and recently settled Ariaal men of northern Kenya. (2008) BMC Evolutionary Biology 8, 173. *Study linking DRD4 polymorphisms to nomadic life in Ariaals of Kenya.*

LaHoste GJ, Swanson JM, Wigal SB, Glabe C, Wigal T, King N, Kennedy JL: Dopamine D4 receptor gene polymorphism is associated with attention deficit hyperactivity disorder. (1996). Mol Psychiatry 1, 121–124. *Original study linking DRD4 polymorphisms with ADHD.*

Settle JE, Dawes CT, Christakis NA, Fowler JH. Friendships Moderate an Association between a Dopamine Gene Variant and Political Ideology. (2010). The Journal of Politics 72, 1189–1198. *In this study an association was reported between political ideology and DRD4 polymorphisms.*

Christakis NA and Fowler JH. Connected: The Surprising Power of Our Social Networks and How They Shape Our Lives—How Your Friends' Friends' Friends Affect Everything You Feel, Think, and Do. Reprint edition. (2011). Back Bay Books. 368 pages. *The book authored by the investigators of the previous study exploring the consequences of human networking in our daily lives.*

Dreber, A, CL Apicella, DTA Eisenberg, JR Garcia, R Zamore, JK Lum & BC Campbell. The 7R Polymorphism in the Dopamine Receptor D4 Gene (DRD4) is associated with financial risk-taking in men. (2009). Evolution and Human Behavior, 30, 85–92.

Kuhnen CM and Chiao JY Genetic Determinants of Financial Risk Taking. (2009). PLoS ONE 4(2): e4362. doi:10.1371/journal.pone.0004362. *These two studies link DRD4 polymorphisms with financial risk taking decisions.*

183

Weber EU and Hsee CK. Culture and individual judgment and decision making. (2000). Applied Psychology: An International Review 49, 32–61. *In this comprehensive review article the authors discuss and cite several studies addressing the role of cultural differences in decision making.*

Hsee CK and Weber CK. Cross-national differences in risk preference and lay predictions. (1999). Journal of Behavioral Decision Making 12, 165–179.

Weber CK and Hsee CK. Cross cultural differences in risk perception but cross cultural similarities in attitude towards perceived risk. (1998). Management Science 44, 1205–1217. *These articles explore differences between Chinese and Americans in the perception of risk and the subsequent decisions.*

Lau LY. Chinese and English Probabilistic Thinking and Risk Taking in Gambling. (2005). Journal of Cross-Cultural Psychology. 36, 621–627. *Comparison in gambling behavior between Chinese and Westerners.*

Pitta, D.A., Fung, H.G., & Isberg, S. Ethical issues across cultures: Managing the differing perspectives of China and the USA. (1999). Journal of Consumer Marketing, 16, 240–256. *Study describing how Chinese view success.*

Rohrmann B and Chen H. Risk perception in China and Australia: an exploratory crosscultural study. (1999). Journal of Risk Research 2, 219–241. *This study addresses the perception of risk in China and Australia.*

Lusher JM, Chandler C and Ball D. Dopamine D4 receptor gene (DRD4) is associated with Novelty Seeking (NS) and substance abuse: the saga continues … (2001). Molecular Psychiatry 6, 497–499. *Critically reviews the results non DRD4 and drug addiction and abuse.*

Chapter 8

Goldman N, Glei DA, Lin Y-H. The Serotonin Transporter Polymorphism (5-HTTLPR): Allelic Variation and Links with Depressive Symptoms. (2010). Depress Anxiety. 27, 260–269. *Provides a detailed and recent account of results exploring the frequency of 5-HTTLPR in different ethnic groups.*

Murakami F, Shimomura T, Kotani K, Ikawa S, Nanba E, Adachi K. Anxiety traits associated with a polymorphism in the serotonin transporter gene regulatory region in the Japanese. J Hum Genet. 1999;44(1):15–7. *One of the earliest discussions of the difference between the 5-HTTLPR allelic frequencies between Asians and Caucasians.*

Caspi A, Sugden K, Moffitt TE, Taylor A, Craig IW, Harrington H, McClay J, Mill J, Martin J, Braithwaite A, Poulton R. Influence of life stress on depression: moderation by a polymorphism in the 5-HTT gene. (2003). Science. 301, 386–389.

Karg K, Burmeister M, Shedden K, Sen S. The serotonin transporter promoter variant (5-HTTLPR), stress, and depression meta-analysis revisited: evidence of genetic moderation. (2011). Arch Gen Psychiatry. 68, 444–454. *These studies exemplify the analyses linking the presence of the s allele of 5-HTT with the manifestation of depressive symptoms.*

Taylor SE, Way BM, Welch WT, Hilmert CJ, Lehman BJ, Eisenberger NI. Early family environment, current adversity, the serotonin transporter promoter polymorphism, and depressive symptomatology. (2006). Biol Psychiatry. 60, 671–676. *This study showed that when ss individuals were exposed to positive events they had less depressive symptomatology than those with sl and ll individuals.*

Kilpatrick DG, Koenen KC, Ruggiero KJ, Acierno R, Galea S, Resnick HS, Roitzsch J, Boyle J, Gelernter J. The serotonin transporter genotype and social support and moderation of posttraumatic stress disorder and depression in hurricane-exposed adults. (2007). Am J Psychiatry. 164, 1693–1699. *Study showing that the s allele carriers are more sensitive to social support than the l allele carriers.*

Chiao JY, Blizinsky KD. Culture–gene coevolution of individualism–collectivism and the serotonin transporter gene. (2010). Proc. R. Soc. B 277, 529–537.

Way BM and Lieberman MD (2010). Is there a genetic contribution to cultural differences? Collectivism, individualism and genetic markers of social sensitivity. SCAN (2010) 5, 203–211. *These excellent reviews meticulously describe the genetic evidence that link the s allele of the serotonin transporter with collectivistic cultures and behavior.*

Hofstede, G. Culture's consequences: comparing values, behaviors, institutions and organizations across nations. (2001). Thousand Oaks, CA: Sage Publications. *This book offers a description and geographic distribution of individualistic and collectivistic behavior and cultures across various different regions of the world.*

Fincher, C. L., Thornhill, R., Murray, D. R. & Schaller, M. Pathogen prevalence predicts human cross-cultural variability in individualism/collectivism. (2008) Proc. R. Soc. B 275, 1279–1285. *Study describing the protective effects of collectivistic cultures against pathogen prevalence.*

Chang, D.F. Understanding the rates and distribution of mental disorders. In: Kurasaki, K.S., Okazaki, S., Sue, S., editors. Asian American Mental Health: Assessments Theories and Methods. (2002). New York: Kluwer Academic, pp. 9–27.

Hwang, W., Myers, H. Major depression in Chinese Americans. (2007). Social Psychiatry and Psychiatric Epidemiology 42, 189–197. *These studies describe that*

East Asians living in the US exhibit more frequently symptoms of depression than East Asians living in Asia, in societies with collectivistic cultures.

De Neve, J-E, Christakis, NA., Fowler, JH. and Frey, BS., Genes, Economics, and Happiness (2010). CESifo Working Paper Series No. 2946. Available at SSRN: http://ssrn.com/ abstract=1553633. *Association between 5-HTT polymorphisms and happiness.*

Szekely A, Ronai Z, Nemoda Z, Kolmann G, Gervai J, Sasvari-Szekely M. Human personality dimensions of persistence and harm avoidance associated with DRD4 and 5-HTTLPR polymorphisms (2004). Am. J. Med. Genet. B Neuropsychiatr. Genet. 126B, 106–110.

Auerbach JG, Faroy M, Ebstein R,Kahana M, Levine L The Association of the Dopamine D4 Receptor Gene (DRD4) and the Serotonin Transporter Promoter Gene (5-HTTLPR) with Temperament in 12- month-old Infants. (2001). Journal of Child Psychology and Psychiatry 42, 777–783.

Hohmann S, Becker K, Fellinger J, Banaschewski T, Schmidt MH, Esser G, Laucht M. Evidence for epistasis between the 5-HTTLPR and the dopamine D4 receptor polymorphisms in externalizing behavior among 15-year-olds. (2009). J Neural Transm. 116, 621–629.

Auerbach JM, Geller V, Lezer S, Shinwell E, Belmaker RH, Levine J, Ebstein R Dopamine D4 receptor (D4DR) and serotonin transporter promoter (5-HTTLPR) polymorphisms in the determination of temperament in 2-month-old infants. (1999). Molecular Psychiatry 4, 369–373.

Lakatos K, Nemoda Z, Birkas E, Ronai Z, Kovacs E, Ney K, Toth I, Sasvari-Szekely M, Gervai J. Association of D4 dopamine receptor gene and serotonin transporter promoter polymorphisms with infants' response to novelty. (2003). Mol Psychiatry. 8, 90–97. *Studies that address the combined effects of DRD4 and 5-HTT polymorphisms in behavior.*

Stallings, M. C., Hewitt, J. K., Cloninger, C. R., Heath, A. C., & Eaves, L. J. Genetic and Environmental Structure of the Tridimensional Personality Questionnaire: Three or Four Temperament Dimensions? (1996). Journal of Personality and Social Psychology, 70, 127–140. *Description of the four-factor model for personality dimensions, according to the personality theorist C.R. Cloninger.*

Pai C-Y, Chou S-L, Huang F,F-Y. Assessment of the role of a functional VNTR polymorphism in MAOA gene promoter: a preliminary study. (2007). Forensic Science Journal 6, 37–43.

Sabol S.Z., Hu S. and Hamer D. A functional polymorphism in the monoamine oxidase A gene promoter. (1998). Human Genetics 103, 273–279.

Jorm AF, Henderson AS, Jacomb PA, Christensen H, Korten AE, Rodgers B, Tan X, Easteal S. Association of a functional polymorphism of the monoamine oxidase A gene promoter with personality and psychiatric symptoms. (2000). Psychiatr Genet 10, 87–90.

Deckert J, Catalano M, Syagailo YV, Bosi M, Okladnova O, Di Bella D, Nothen MM, Maffei P, Franke P, Fritze J, Maier W, Propping P, Beckmann H, Bellodi L, Lesch KP. Excess of high activity monoamine oxidase A gene promoter alleles in female patients with panic disorder. (1999). Hum Mol Genet 8, 621–624.

Hamilton SP, Slager SL, Heiman GA, Haghighi F, Klein DF, Hodge SE, Weissman MM, Fyer AJ, Knowles JA. No genetic linkage or association between a functional promoter polymorphism in the monoamine oxidase-A gene and panic disorder. (2000). Mol Psychiatry 5, 465–466. *These studies report the incidence of MAOA uVNTR in various ethnic groups.*

Kim-Cohen, J., Caspi, A., Taylor, A., et al.. MAOA, maltreatment, and gene-environment interaction predicting children's mental health: new evidence and a meta-analysis. (2006). Molecular Psychiatry, 11, 903–913.

Widom, C.S., Brzustowicz, L.M. MAOA and the "cycle of violence:" Childhood abuse and neglect, MAOA genotype, and risk for violent and antisocial behavior. (2006). Biological Psychiatry, 60, 684–689.

Ducci, F., Enoch, M.A., Hodgkinson, C., et al. Interaction between a functional MAOA locus and childhood sexual abuse predicts alcoholism and antisocial personality disorder in adult women. (2007). Molecular Psychiatry, 13, 334–347.

Caspi A, McClay J, Moffitt TE, Mill J, Martin J, Craig IW, et al. Role of genotype in the cycle of violence in maltreated children. (2002). Science 297, 851–854. *These studies associate MAOA-uVNTR with response to certain environmental stimuli.*

Eisenberger NI, Way BM, Taylor SE, Welch WT, Lieberman MD. Understanding Genetic Risk for Aggression: Clues From the Brain's Response to Social Exclusion. (2007). Biological Psychiatry 61, 1100–1108.

Cases O, Seif I, Grimsby J, Gaspar P, Chen K, Pournin S, et al. Aggressive behavior and altered amounts of brain serotonin and norepinephrine in mice lacking MAOA. (1995). Science 268:1763–1766. *Evidence between MAOA-uVNTR polymorphisms and aggression in humans and mice.*

Lea R and Chambers G. Monoamine oxidase, addiction, and the "warrior" gene. hypothesis. (2007). THE NEW ZEALAND MEDICAL JOURNAL 120 No

1250 ISSN 1175 8716. *Discussion on the warrior gene findings on Maori people of New Zeland.*

McDermott R, Tingley D, Cowden J, Frazzetto G and Johnsone DDP. Monoamine oxidase A gene (MAOA) predicts behavioral aggression following provocation.

Chapter 9

Auton, A. et al. A global reference for human genetic variation. Nature 526, 68–74 (2015). https://doi.org/10.1038/nature15393 *The 1000 Genomes Project Consortium.*

Chen, F.S., R. Kumsta, B. von Dawans, M. Monakhov, R.P. Ebstein, M. Heinrichs. Common oxytocin receptor gene (OXTR) polymorphism and social support interact to reduce stress in humans. Proc. Natl. Acad. Sci. U. S. A., 108 (2011), pp. 19937–19942. *Variaions in oxytocin receptor and stress.*

Karayiorgou M, Altemus M, Galke BL, Goldman D, Murphy DL, Ott J, Gogos JA. Genotype determining low catechol-O-methyltransferase activity as a risk factor for obsessive-compulsive disorder. (1997). Proc Natl Acad Sci U.S.A. 94, 4572–4575.

Kotler M, Barak P, Cohen H, Averbuch IE, Grinshpoon A, Gritsenko I, Nemanov L, Ebstein RP. Homicidal behavior in schizophrenia associated with a genetic polymorphism determining low catechol O-methyltransferase (COMT) activity. (1999). Am J Med Genet. 88, 628–633.*These study exemplify the association of the low activity A allele for COMT with certain psychopathologies and neurological conditions.*

Kunugi H, Nanko S, Ueki A, Otsuka E, Hattori M, Hoda F, Vallada HP, Arranz MJ, Collier DA. High and low activity alleles of catechol-O-methyltransferase gene: ethnic difference and possible association with Parkinson's disease. (1997). Neurosci Lett. 221, 202–204.

Lee Y-T, Norasakkunkit V, Liu L, Zhang J-X, Zhou M-J. Daoist/Taoist Altruism and Wateristic Personality: East and West. (2008). World Cultures eJournal, 16(2). *Cross-cultural comparison of altruistic tendency, between Americans and Chinese.*

Li, J., Zhao, Y., Li, R., Broster, L. S., Zhou, C., and Yang, S. (2015). association of oxytocin receptor gene (OXTR) rs53576 polymorphism with sociality: a meta-analysis. PLoS One 10:e0131820. doi: 10.1371/journal.pone.0131820. *Association between oxytocin receptor polymorphisms and optimism.*

Palmatier MA, Min Kang A, and Kidd KK. Global Variation in the Frequencies of Functionally Different Catechol-O-Methyltransferase Alleles. (1999).

Biol Psychiatry. 46, 557–567. *Survey of the allelic frequencies of COMT in different populations around the world.*

Reuter M, Frenzel C, Walter NT, Markett S, Montag C. Investigating the genetic basis of altruism: the role of the COMT Val158Met polymorphism. (2011) Soc Cogn Affect Neurosci. 6, 662–668. *A study demonstrating that the Val-containing COMT allele is associated with altruistic behavior. The study was based on the assessment of the amount of money donated to a poor child in a developing country.*

Stein DJ, Newman TK, Savitz J, Ramesar R. Warriors versus worriers: the role of COMT gene variants. (2006). CNS Spectr. 11, 745–738. *Discussion of the worrier vs. warrior strategies in association with COMT.*

Walter Bradford Cannon. Bodily changes in pain, hunger, fear, and rage. (1929). New York: Appleton-Century-Crofts. *One of the earliest descriptions of the" fight or flight" response.*

Chapter 10

Bierut L.J., *et al.* Novel genes identified in a high-density genome wide association study for nicotine dependence. (2007). Hum. Mol. Genet. 16, 24–35.

Saccone S.F. *et al.* Cholinergic nicotinic receptor genes implicated in a nicotine dependence association study targeting 348 candidate genes with 3713 SNPs. (2007). Hum. Mol. Genet. 16, 36–49.

Zeiger *et al.* The neuronal nicotinic receptor subunit genes (CHRNA6 and CHRNB3) are associated with subjective responses to tobacco. (2008). Hum. Mol. Genet. 17, 724–734. *CHRNB3 and tobacco dependence.*

De Neve JE, Mikhaylov M, Dawes CT, Christakis N, Fowler JH. Born to Lead? A Twin Design and Genetic Association Study of Leadership. (August 24, 2011). SSRN: http://ssrn.com/abstract=1915939

Chapter 11

Rietveld CA, Medland SE, Derringer J, *et al.* GWAS of 126,559 individuals identifies genetic variants associated with educational attainment. Science. 2013 Jun 21;340(6139):1467–71. doi: 10.1126/science

van den Berg SM, de Moor MH, Verweij KJ, *et al.* Meta-analysis of Genome-Wide Association Studies for Extraversion: Findings from the Genetics of Personality Consortium. Behav Genet. 2016 Mar;46(2):170–82. doi: 10.1007/s10519-015-9735-5

Terracciano A, Esko T, Sutin AR, et al. Meta-analysis of genome-wide association studies identifies common variants in CTNNA2 associated with excitement-seeking. Transl Psychiatry. 2011;1:e49. https://doi.org/10.1038/tp.2011.42

Sanchez-Roige S, Gray JC, MacKillop J, Chen CH, Palmer AA. The genetics of human personality. Genes Brain Behav. 2018 Mar;17(3):e12439. doi: 10.1111/gbb.12439. Epub 2017 Dec 29. PMID: 29152902; PMCID: PMC7012279.

Lo MT, Hinds DA, Tung JY, et al. Genome-wide analyses for personality traits identify six genomic loci and show correlations with psychiatric disorders. Nat Genet. 2017;49(1):152-156. doi:10.1038/ng.3736

Boutwell B, Hinds D; 23andMe Research Team, Tielbeek J, Ong KK, Day FR, Perry JRB. Replication and characterization of CADM2 and MSRA genes on human behavior. Heliyon. 2017 Jul 26;3(7):e00349. doi: 10.1016/j.heliyon.2017.e00349. PMID: 28795158; PMCID: PMC5537199.

Day FR, Helgason H, Chasman DI, Rose LM, Loh P-R, Scott RA, Helgason A, Kong A, Masson G, Magnusson OT, et al. Physical and neurobehavioral determinants of reproductive onset and success. Nat. Genet. (2016), pp. 617–623.

Kim HS, Sherman DK, Sasaki JY, Xu J, Chu TQ, Ryu C, Suh EM, Graham K, Taylor SE. Culture, distress, and oxytocin receptor polymorphism (OXTR) interact to influence emotional support seeking. Proc Natl Acad Sci U S A. 2010 Sep 7;107(36):15717–21. doi: 10.1073/pnas.1010830107. Epub 2010 Aug 19. PMID: 20724662; PMCID: PMC2936623.

de Moor MHM, Costa PT, Terracciano A, et al. Meta-analysis of genome-wide association studies for personality. Mol Psychiatry. 2012;17:337–349. https://doi.org/10.1038/mp.2010.128.

Kim H, Roh S-J, Sung YA, et al. Genome-wide association study of the five-factor model of personality in young Korean women. J Hum Genet. 2013;58:667–674. https://doi.org/10.1038/jhg.2013.75

Hill MN, Patel S. Translational evidence for the involvement of the endocannabinoid system in stress-related psychiatric illnesses. Biol Mood Anxiety Disord. 2013 Oct 22;3(1):19. doi: 10.1186/2045-5380-3-19. PMID: 24286185; PMCID: PMC3817535.

De Neve JE, Mikhaylov S, Dawes CT, Christakis NA, Fowler JH. Born to Lead? A Twin Design and Genetic Association Study of Leadership Role Occupancy. Leadersh Q. 2013;24(1):45–60. doi:10.1016/j.leaqua.2012.08.001

Bae HT, Sebastiani P, Sun JX, et al. Genome-wide association study of personality traits in the long life family study. Front Genet. 2013;4:65.

Okuyama Y, Ishiguro H, Nankai M, Shibuya H, Watanabe A & Arinami T. (January 2000). "Identification of a polymorphism in the promoter region of

DRD4 associated with the human novelty seeking personality trait". Molecular Psychiatry. 5 (1): 64–69. doi:10.1038/sj.mp.4000563. PMID 10673770.

Ronai Z, Szekely A, Nemoda Z, Lakatos K, Gervai J, Staub M & Sasvari-Szekely M. (January 2001). "Association between Novelty Seeking and the -521 C/T polymorphism in the promoter region of the DRD4 gene". Molecular Psychiatry. 6(1): 35–38. doi:10.1038/sj.mp.4000832. PMID 11244482.

Söderqvist S, Matsson H, Peyrard-Janvid M, Kere J, Klingberg T. Polymorphisms in the dopamine receptor 2 gene region influence improvements during working memory training in children and adolescents. J Cogn Neurosci. 2014 Jan;26(1):54–62. doi: 10.1162/jocn_a_00478. Epub 2013 Sep 3. PMID: 24001007.

Montag C, Jurkiewicz M, Reuter M. The role of the catechol-O-methyltransferase (COMT) gene in personality and related psychopathological disorders. CNS Neurol Disord Drug Targets. 2012;11(3):236–250. doi:10.2174/187152712800672382

Hashimoto R, Noguchi H, Hori H, Ohi K, Yasuda Y, Takeda M, Kunugi H. A possible association between the Val158Met polymorphism of the catechol-O-methyl transferase gene and the personality trait of harm avoidance in Japanese healthy subjects. Neurosci. Lett. 2007;428(1):17–20.

Schmitt DP, Allik J, McCrae RR, Benet-Martínez V. The Geographic Distribution of Big Five Personality Traits: Patterns and Profiles of Human Self-Description Across 56 Nations. Journal of Cross-Cultural Psychology. 2007;38(2):173–212. doi:10.1177/0022022106297299

Fernandes, D.M., Sirak, K.A., Ringbauer, H. et al. A genetic history of the pre-contact Caribbean. Nature (2020). https://doi.org/10.1038/s41586-020-03053-2

Nägele, K. et al. Genomic insights into the early peopling of the Caribbean. Science 369, 456–460 (2020).

These two recent studies analyzed genomic data from populations in the Caribbean. The former did not, but the latter did infer the occurrence of early migration events from North America. The major portion though of migrations that established the populations of the Caribbean, apparently had originated from Central and Northeastern South America.

Chapter 12

Papagrigorakis MJ, Yapijakis C, Synodinos PN, Baziotopoulou-Valavani E. DNA examination of ancient dental pulp incriminates typhoid fever as a probable cause of the Plague of Athens. (2006). International Journal of Infectious

Diseases 10, 206–214. *Identification of typhoid fever as a cause of the Plague of Athens by DNA analysis.*

Jackson T. Prosperity without Growth: Economics for a Finite Planet (2011). Routledge. 2011 pages. *An interesting critique of the existing economic paradigm of continuous growth and individualism.*

Chapter 13

Levitt SD and Dubner SJ. Super Freakonomics: Global Cooling, Patriotic Prostitutes, and Why Suicide Bombers Should Buy Life Insurance. (2009). William Morrow, 288 pages. *Among other interesting and frequently surprising observations discusses the immediate influences of Western television series in birth control in India. This issue was unsuccessfully attempted to be solved by various legal regulations prior to the introduction of television.*

McLuhan M. The Gutenberg Galaxy: The Making of Typographic Man. (1962). University of Toronto Press. 293 pages. *With this book the author has introduced the term Global Village. He has focused on the role of mass communication in culture and human affairs.*

Fukuyama F. The End of History and the Last Man. (1993). Harper Perennial. 448 pages. *The controversial book that predicts that the Western liberal democracy may signal the end point of humanity's socio-cultural evolution and the final form of human government in a state that more or less will be universal and homogenous.*

Chapter 14

Isaac Asimov. Foundation (1991). Spectra Publishers. 320 pages. *This is the first book of the Foundation series, originally published in 1951 (Gnome Press). Other books of the Foundation series include Foundation and Empire (1952), Second Foundation (1953), Foundation's Edge (1982), Foundation and Earth (1986) and others. In this book the author has introduced the concept of psychohistory, a science that facilitates the general prediction of future behaviors of very large groups of people. This hypothetical science is based on history, sociology, and mathematical statistics.*

Notes

1. Castoriades Cornelius (1964) in Anthology Socialisme ou Barbarie' - No. 35. Published under pseudonym Paul Carden in "Redefining Revolution" para. 29. Several English translations are available such as the one from Cornell University Press No. 3 1224 087 844 753

2. *The phrase "nature or nurture" returns more than 16 x 10⁶ hits in a Google search, which exemplifies the intensity of this debate.* It becomes even more complicated when a sharp borderline between the two – or the self and the environment- in certain cases cannot be defined. In that sense, occasionally the individual (self) by being part of the environment manipulates it and thus, strictly speaking, the latter should not be considered as an extrinsic parameter.

3. This is again an oversimplification because certain polymorphisms may benefit, instead of the individual, a group of individuals or the population. Such phenomena are classified under the term of "group selection" and are discussed in the textbooks on evolutionary biology cited earlier. In that case, the essential issue is related to the identification of what the actual subject of evolution is, namely the individual, the group or the genes.

4. The web site of "International Commission on Zoological Nomenclature" is http://iczn.org/category/faqs/frequently-asked-questions/who-type-homo-sapiens and at this one can find the official name of all animals. In this site (http://taxonomicon.taxonomy.nl/TaxonList.aspx?subject=Taxon&by=-ScientificName&search=homo+sapiens*,) variations of the human species according to Linnaeus are being described.

5. The classification of *Homo monstrosus* as a distinct species from *Homo sapiens* has also been proposed by *Linneaus*.

6. Two-feet standing or *bipedalism* has not developed only in apes but at other times during evolution, such as in macropods that include the kangaroos. Bipedal dinosaurs have also occurred and the birds are the evolutionary descendents of these now extinct species.

7. This "go together" tendency for certain behavioral and other traits is different from the genetic linkage that is due to the fact that 2 polymorphic genes are physically close in the chromosomes. In that case, the corresponding traits co-segregate in the offspring. A measurement of this genetic distance is obtained by assessing the degree of recombination between these alleles which reflects the frequency by which the traits do not segregate together in the offspring. In the "go together" case we have traits that are not genetically linked, but they have the tendency to co-exist in the same population. This is simply due to the fact that the corresponding genes (alleles) have higher frequency in the same population, and therefore the traits associated with them appear more frequently together than sperately.

8. Meme: an idea, behavior, style, or usage that spreads from person to person within a culture (source: Merriam-Webster Dictionary).

9. This term refers to a set of genetic markers that due to their proximity tend to be inherited together. It may refer to either genomic (in chromosomes) or mitochondrial markers.

10. The exact cause of death of Alexander the Great in Babylon, in 323 BC remains inconclusive. Among various potential causes, the usual suspects that included malaria and typhoid fever have been suggested, along with others such as pancreatitis, West Nile virus infection, syphilis and even assassination.

11. A fire in Oldsmobile manufacturing facility in 1901 destroyed all but one prototypes. Among them luxurious or other exotic cars were included. The one that survived the fire was a so called *Curved Dash* that was placed near the exit and was developed with the intension to target the average American consumer that needed a simple and cheap car. The manufacturing and commercial future of Oldsmobile was shaped by this prototype that survived the fire and the legacy was continued even after its purchase of Oldsmobile from General Motors in 1929 (actually being part of the General Motors' companion make program) and was making relatively cheap cars for the average American.

12. Naturally, direct comparison between individuals and populations is like comparing eggs and apples. The meaning here is that by knowing the genotype of an individual it is possible to compare it with certain populations' allelic frequencies and extract conclusions regarding how common or uncommon the corresponding individual's alleles are. This information can be interpreted as a similarity to this specific population.

13. In 2009 a Pilot Project was initiated in order to distinguish Somali asylum seekers from other Africa nations, based on the results of DNA analyses. Following criticism from many directions the Project was cancelled in 2011. (http://blogs.nature.com/news/2011/06/uk_immigration_cancels_dna_scr_1.html)

14. Direct to consumers DNA tests refer to the DNA tests that are accessible directly to the consumer without having to go through a health care professional. Several controversies have been triggered regarding these tests, in principle related to the possible misinterpretation of the results by the consumers. In 2010, the *Walgreens* pharmacy store in the U.S. postponed providing genetic tests through its drugstores after the Food and Drug Administration challenged the legality of the test. Over time such issues are being resolved but the controversy regarding their utility persists. http://www.nytimes.com/2010/05/13/health/13gene.html

15. Representative and widely known sayings of Heraclitus are the: *"No man ever steps in the same river twice, for it's not the same river and he's not the same man"* and *"Everything changes, and nothing abides".*

16. The discussion about the true motives of religious wars starting from the Crusades during the Middle Ages, until the religious wars in Modern and contemporary societies is apparently a matter of debate and depending on the specific view-point various different deeper motives will be identified. However, undoubtedly a great role was also played by the strong belief that they served a greater cause, for which religious rightness was instrumental, providing to the least a moral excuse.

17. The following link provides a lively presentation of the "6 killer Western applications" by the author Niall Ferguson. http://macromon.wordpress.com/2011/11/04/niall-ferguson-the-6-killer-apps-of-prosperity/

18. An interesting article by M. Brooks that appeared in the New Statement in 16/08/2010 and entitled "The spark rises in the east" discusses the recent investment of China to science. http://www.newstatesman.com/asia/2010/08/china-research-chinese-science. Among other issues, in this interesting article, the fact that hierarchical Eastern traditions have to give way to more liberal practices is proposed, as a perquisite for the success of this effort. Noteworthy, the rigid hierarchal structure is integral to the collectivistic societies.

19. It could be argued that various remarkable accomplishments and human artifacts of these Central and South American civilizations, such as the

Aztec's calendar sunstone and the Mayan astronomy may indeed indicate evidence for scientific advancement. However, in our case we focus on the curiosity-driven science that is based in logic and rational. Probably, future archeological research will better understand the essence of these accomplishments and categorize them as evidence of scientific progress, which in turn may resolve this phenomenological discrepancy.

20. President's Obama statement of 01/21/2010 on economic crisis can be found in: http://www.ft.com/cms/s/0/26e0f512-06b4-11df-b426-00144fe-abdc0.html#axzz1Yaht FSLY

21. *Harm avoidance* is a behavioral trait for which individuals with high score are described as worrying and pessimistic, fearful and doubtful, shy, and fatigable. The personality theorist R.C. Cloninger (Stallings et al, 1996) proposed that *harm avoidance*, along with *novelty seeking, persistence*, and *reward dependence* constitute the four temperamental dimensions. It is noted that in this classification, persistence was added at a subsequent point as soon as it was realized that it represents an independent factor.

22. *Surui, Karitiana* and the *Ticuna* are indigenous populations of South America. The population of the Suruis today is about 1000, of the Karitianas about 300 while of the Ticuna Indians about 35000 individuals. The reservations for all of these people are in the Amazon rainforest.

23. The HapMap project was a large scale worldwide genetic project that records and studies the genetic variation among different world populations (http://hapmap.ncbi.nlm.nih.gov/index.html.en). It was replaced by the 1000 Genomes Project that collects variation data from different populations worldwide.

24. Taxation data were downloaded from https://tradingeconomics.com and data on economic growth and sustainability from the World bank portal (https://databank.worldbank.org/home.aspx). For both, current data as of November 2020 were used.

25. The World Governance Indicators were developed by Daniel Kaufmann (Natural Resource Governance Institute (NRGI) and Brookings) and Aart Kraay, (World Bank, Development Economics) and can be retrieved by http://info.worldbank.org/governance/wgi/

26. These are only some of the ethnic groups found in Africa. Yoruba people are about 30 millions and are found to West Africa. Maasai and Luhya are found in Kenya and Tanzania and are estimated to about 1.5 millions and 6.5 millions respectively.

27. Hierarchical clustering analysis for this example was performed using the publicly available Morpheus analysis software from Broad Institute (MIT, Boston, MA) that can be found by using this link: (https://software. broadinstitute.org/morpheus). Several such tools are available.

28. In theory, having DNA samples from certain populations representing different points of historical time may have allowed recording of such periodic variation of allelic frequencies. However, the major limitation is that population migration phenomena are quite likely to have masked the potential periodic changes in allelic frequencies. In other words, we would not be able to interpret s alleles' increases to the advantage of s-allele bearing individuals or to the migration of s-allele bearing individuals into this population. Of course, by monitoring allelic frequencies in other genetic loci and by interpreting them in combination with the frequency of s allele we could infer whether s allele's frequency changes are due to migrations.

29. The term *"butterfly effect"* was coined by the mathematician and meteorologist Edward Norton Lorenz in order to describe that a small change at one place in a nonlinear system can result in large differences to a later state. The illustrative analogy is related to the flapping of the wings of a butterfly that can cause hurricane elsewhere.

30. Certain groups of people have lived in isolation for considerable time periods. Aboriginals in Australia in example are thought to represent the oldest group of people outside Africa that resembles genetically the original pre-historic settlers of this area. These groups and populations though, while devoid to considerable gene exchanges from outside populations, they were quite limited arithmetically so they could not account for the development of cultures that have emerged to defined civilizations, at least of the magnitude of others.

31. Non-random mating between individuals of the same population results in systematic differences in the allelic frequencies between subpopulations, a phenomenon designated as *population stratification*. It can be due to physical separation, the existence of cultural barriers and all other factors that may induce non-random mating. It is eliminated after prolonged periods of non-selective mating.

32. Issues related to the creation of an open national identity, a perquisite for smooth and efficient cultural homogenization, especially in view of currently occurring immigration and how it was dealt in Europe as compared to U.S.A. and Canada are being discussed by Fukuyama in "Identity, immigration

and liberal democracy" that appeared in the *"Journal of Democracy"* 17, 2, April 2006.

33. The observer effect describes the fact that the actual act of observation disturbs the system and eventually influences the outcome of the experiment or phenomenon observed. It is originally described in physics but apparently is relevant to all disciplines.

34. It seems that during and after the COVID-19 crisis China's economy not only regained its growth pace earlier than the U.S. but also will continue to grow. (Reuters: Analysis: China and U.S. economies diverge over coronavirus response, https://www.reuters.com/article/health-coronavirus-usa-china-analysis-idUSKBN277066)

35. In March 2011 an earthquake caused a big Tsunami that inactivated the cooling systems in three Fukushima nuclear reactors. This caused a nuclear accident. 2259 disaster-related deaths were reported and 100,000 were evacuated from their homes. https://www.world-nuclear.org/information-library/safety-and-security/safety-of-plants/fukushima-daiichi-accident.aspx

36. This refers to the European Sovereign Debt Crisis that started around 2008 and picked in 2010–2012. It was triggered by the collapse of the Iceland's banking system and soon it spread to several southern European countries (Portugal, Italy, Ireland, Greece, and Spain) that were unable to refinance (or repay) their government debt. It was eventually controlled by the financial guarantees of European countries, and by the International Monetary Fund (IMF) to prevent the collapse of the euro and financial contagion elsewhere.

www.ingramcontent.com/pod-product-compliance
Lightning Source LLC
Chambersburg PA
CBHW071713170526
45165CB00005B/2004